T0133037

Bombs Away

Bombs Away

Militarization, Conservation, and Ecological Restoration

DAVID G. HAVLICK

The University of Chicago Press
Chicago and London

The University of Chicago Press, Chicago 60637
The University of Chicago Press, Ltd., London

Published 2018
Printed in the United States of America

27 26 25 24 23 22 21 20 19 18   1 2 3 4 5

ISBN-13: 978-0-226-54754-1 (cloth)
ISBN-13: 978-0-226-54768-8 (e-book)
DOI: https://doi.org/10.7208/chicago/9780226547688.001.0001

Library of Congress Cataloging-in-Publication Data
Names: Havlick, David G., author.
Title: Bombs away : militarization, conservation, and ecological restoration / David G. Havlick.
Description: Chicago : The University of Chicago Press, 2018. | Includes bibliographical references and index.
Identifiers: LCCN 2017056178 | ISBN 9780226547541 (cloth : alk. paper) | ISBN 9780226547688 (e-book)
Subjects: LCSH: Restoration ecology—United States. | Military base conversion—United States. | Nature conservation—United States. | Restoration ecology. | Military base conversion. | Nature conservation.
Classification: LCC QH541.15.R45 H38 2018 | DDC 333.73/153—dc23
LC record available at https://lccn.loc.gov/2017056178

♾ This paper meets the requirements of ANSI/NISO Z39.48-1992 (Permanence of Paper).

For Marion

CONTENTS

ONE

Military Natures

On the December morning when I first visited Big Oaks National Wildlife Refuge (NWR), an aging crust of snow covered the southern Indiana ground. With the refuge's deer and turkey hunting seasons already past, I was the day's only visitor. I sat alone in the garage of the refuge office to watch a thirty-minute video. It was not a Discovery Channel thriller. In place of scenic vistas or alluring scenes of wildlife common to many visitor center productions, the film offered images of rusted bomb casings and military spotting charges to help me identify unexploded ordnance (UXO).[1] The video also provided a brief account of decades of munitions testing that took place at the site prior to its designation as a national wildlife refuge.

The refuge was calm on this particular day, but for years the site of Big Oaks shuddered with the sounds of mortar shells. From 1942 to 1992, this was the US Army's Jefferson Proving Ground, a munitions testing facility that as a wildlife refuge remains cluttered with millions of rounds of UXO, twenty-four thousand pounds of depleted uranium, and also, astonishingly, some of the finest songbird, river otter, bobcat, and Indiana bat habitat found anywhere in the midwestern United States. Here—as in a growing number of wildlife refuges in the United States and many other militarized landscapes around the world—there is an unlikely convergence of military activity and environmental protection; the relationship between these two is fraught with complexity that carries both opportunity and risk. The bombs are mostly quiet now at Big Oaks, but they continue to shape the landscape through a curious mix of ecological recovery and human impact.

One of the more surprising and extensive land use changes in the United States since the late 1980s is the conversion of military lands to new classifications as national wildlife refuges. During this time, roughly two dozen military sites have closed or "realigned" to focus instead on wildlife conservation. Similar transitions are occurring—both formally and informally—at sites around the world previously known primarily for their violence or militarization. The Iron Curtain borderlands of Central Europe, for example, now feature in redemptive stories as the "Green Belt of Europe"; and the resolutely *militarized* "demilitarized zone" (DMZ) of the Korean Peninsula attracts tourists not just as a site of historic interest for its protracted tensions, but also for its "supreme tranquility" as an emergent sanctuary for wildlife.[2]

These examples present unusual but valuable places to consider the relationship between defense activities and environmental protection. As landscapes that in various ways can be considered both militarized and natural, sites of military-to-wildlife conversion emerge not as simple natural or social spaces but as blended sites with natural, social, and technological elements.[3] Throughout this book, I describe a number of these places as they exist in the world and how they are represented. This is not an attempt at a global tour of militarized landscapes; rather, these cases scattered across the United States, Europe, Asia, Australia, and the Pacific highlight an important point: militarized landscapes exist in a wide array of locations in a variety of contexts and conditions. With this in mind, it is critical to understand how these places have come to be the way they are and how they are actually created through a combination of political and scientific narratives, and a mix of more tangible human actions that reshape entire landscapes—not simply by virtue of "nature" taking over.

Beyond providing a descriptive account of an intriguing set of landscapes, the examples I offer illuminate the underlying values surrounding national defense, security, and environmental protection that military-to-wildlife conversions signify. As these places emerge from military obscurity—or in some cases, infamy—to more public expression as sanctuaries for wildlife, they bring with them a diverse set of challenges. How, for example, should wildlife refuge personnel trained in wildlife or game management respond to new refuge lands that may be contaminated with UXO, radiation, or chemical weapons? If these sites are too dangerous to accommodate a visiting public, should we consider them "public lands," or for that matter, safe haven for wildlife? And more broadly, as places characterized by both military impacts and environmental protection, how might military-to-wildlife landscapes inform a new understanding of nature and society?

From the outset, I should offer a few words of caution and explanation. In contemporary geography and other fields that examine human-environment relationships critically, scholars often shun or use the term *nature* quite warily. For several decades, a variety of books and articles have proclaimed that we have come to the "end of nature," or that we now live in a postnatural world where human impacts dominate and there is no longer (if ever there was) a separate natural domain of the other-than-human.[4] Of course, humans too are animals, so pointing to a separate "nature" calls forth inevitable questions about the human-nature relationship. I still refer to nature not because I am unaware of or unsympathetic to these critiques, but because for many of us the word still represents something we find meaningful—a realm we appreciate in diverse forms, from a stream or stand of trees in an urban park to extensive wildlands far from our cities. This also highlights the idea that "nature" represents a range of conditions, some seemingly distant from human control, others bound closely to it. The *military natures* I focus on in this book are very much blended places, where even sites that fit traditional notions of "nature" or "wilderness" have been shaped by and still contain important elements of human activity. They are, in other words, hybrid places where nature and culture come together in transformative ways.

I first encountered military-to-wildlife transitions in 2001, when my wife was researching fire management on public lands and discovered a plan for a closed military site in Indiana. The new federal managers of this land proposed burning patches of habitat in order to maintain grasslands favored by songbirds. What caught my attention, though, was the seeming paradox that this former army proving ground included not just 50,000 acres of rare bird, bat, and bobcat habitat, but also a depleted uranium firing range and millions of rounds of UXO. The idea of using fire for habitat restoration at an old bombing range unsettled and intrigued me. Why would anyone come up with such a plan?

As it turns out, there are plenty of reasons military-to-wildlife transitions might seem to make sense. This no doubt helps explain why these kinds of land use changes are now found in a variety of sites around the world. For nearly two decades I've traveled around the United States, to the Caribbean, and to parts of Europe and Asia to explore some of these sites firsthand. Of the Department of Defense sites in the mainland United States that have been formally redirected to conservation purposes in recent decades, I've visited all but one, which is closed to visitors due to its uncontained UXO hazards. Along the way, I've also discovered that some of these military-to-wildlife sites were more or less right next door all along—both the Rocky Mountain Arsenal and the Rocky Flats nuclear plant were visible from the house where I grew up. This points to a second important point: even in places we might not initially recognize, militarized landscapes are often very personal. In my travels, I'm often surprised at how many people express personal connections to these places. Some, like me, grew up near weapons production or testing facilities; others worked at these installations, or had parents, grandparents, spouses, aunts, or uncles who did; increasingly, I hear from people—like my colleague who honeymooned in Vieques, Puerto Rico—who have vacationed in places that were profoundly shaped by decades of militarization, but now beckon as "natural" destinations for wildlife viewing or "pristine" beaches and marine environments.

In this book, I consider militarized landscapes to be places that have been substantially impacted by military or defense activities—training lands, for example, as well as bases, defense installations, proving grounds, and security areas or borderlands fortified or enforced by military power. Groundwater or soils contaminated by military activities fall within this definition too, as the chemicals produced or discarded on military installations migrate to impact areas outside the explicit boundaries of military control. War zones also clearly fit, though these mostly fall beyond the scope of this book. As I researched these places, I came to realize that broad shifts in the geographies of national defense and geopolitics have led to a variety of transitioning uses for militarized landscapes. Many of these landscapes have subsequently been dedicated to conservation purposes. While these changes often present legitimate gains for environmental protection and ecological restoration, they also carry serious risks—not only from the physical hazards of chemicals or ordnance that linger, but by erasing important land use histories and the cultural impacts of war and militarization.

The position I take in this book is, at its core, geographical. I argue that it is both possible and critically important to attend to cultural and ecological interests in ways that promote new understandings about militarized landscapes. This approach ought to honor nature and culture to bring these attributes into conversation with one another not as disparate entities, but as interconnected domains. By taking a closer look at military-to-wildlife transitions—the former bombing ranges, weapons plants, training grounds, and militarized borderlands in more detail—I consider how and why these dramatic changes occur, and what to make of the new mix of militarization, conservation, and ecological restoration these places present. Ultimately this can make us more attentive to the important linkages that exist between conservation and militarization, and even more broadly, nature and politics.

These elements at first seem an uncomfortable fit—and in some ways they remain at odds with one another both practically and conceptually—but the cases I present throughout this book illustrate how military activities, conservation goals, and ecological restoration are made to work together to create new kinds of places and new conceptions of place. In ways that are important to understand more fully, this blending of militarization, conservation, and ecological restoration is creating new geographies, by which I mean new landscapes characterized by deeply intertwined social and natural elements.

Changing Militarized Landscapes

In recent decades, military-to-wildlife (M2W) conversions in the United States alone have affected more than one million acres of land and hundreds of millions of acres of marine territory,[5] including a strange palette of some of the country's most contaminated and best preserved habitats. The complexity and difficulty of cleaning up and managing obsolete military facilities offers important insights, regarding these places and elsewhere, into the science, politics, and public understanding of ecological restoration—or the varied practices brought to bear on damaged and degraded sites. Closing military sites and transitioning them explicitly to wildlife conservation purposes also suggests that militarization can change in character. Depending on how these changes are explained or represented can, in turn, advance a particular suite of national priorities and values. The process of military base conversion itself promises to open historically restricted places to new kinds of public attention,[6] even as it brings to the surface new kinds of challenges to wildlife managers and other federal officials. These reconfigured lands also provide new conservation opportunities (such as habitat protection) while raising important questions about the relationship between militarism and environmentalism.

In sites outside the United States, the politics of military-to-wildlife transition often differ markedly, both from one another and from those in the United States, but the changes taking place still call for a critical examination of the military-environment relationship. The greening of Europe's former Iron Curtain, for example, comes from a distinctive array of conditions reaching prior to the Cold War division of Europe. Efforts here incorporate contemporary aspirations to establish a more unified European identity while also protecting the region's history and environment. Other borderlands, such as Korea's DMZ and the United Nations–designated "Green Line" that divides Cyprus, remain unresolved politically but provide examples of resurgent (if also militarized) nature in the interim.[7]

In order to understand military-to-wildlife conversions, it is important to evaluate how these particular landscapes have been created, and how these processes then shape meaning in the new kinds of places that emerge. The first of these hinges in many ways on the stories we tell about these sites, the narratives we create that highlight the political and scientific justification for their conversion. For the second, I consider how these places work as new assemblages of nature and society, and how efforts to restore these sites ecologically—and later, to commemorate them—play a role in our understanding these lands and the processes that created them.

The Nature of Base Redesignations

Prior to 1942, the site of today's Caddo Lake NWR in rural East Texas consisted primarily of cotton fields and sharecropper farms. In the buildup to the Second World War, an ambitious Texas congressman named Lyndon Baines Johnson urged the US Army to establish a new weapons plant near Karnack, Texas, his wife's hometown. Within months, residents of the site were displaced from nearly 8,500 acres in order to make room for the newly established Longhorn Army Ammunition Plant.[8] For the next five and a half decades, Longhorn manufactured a variety of military products, ranging from explosives such as trinitrotoluene (better known as TNT), button bombs, and artillery rounds to incendiary and pyrotechnic munitions to rocket motors and missile propellants. By the late 1980s, some of the products of Longhorn came full circle as the site received the first Pershing missiles to be dismantled as part of the new Soviet-US intermediate-range nuclear arms treaty. By 1997, the facility was declared "excess" by the Department of Defense (DOD), and in October 2000 the US Fish and Wildlife Service (FWS) designated much of the area surrounding the former weapons plant as Caddo Lake NWR.

I first contacted refuge officials about coming to Caddo Lake in 2006, but at the time no public use was allowed and the two refuge staff working at the site had their hands full, dealing with chemical wastes and explosive hazards left behind by the army, and trying to develop a long-term management plan. In September 2009, the refuge opened to the public, and a year and a half later I finally made my way to East Texas to visit.

According to the FWS, Caddo Lake stands out today as one of the richest examples of Texas Piney Woods and mature, flooded bald cypress forest found anywhere in the United States. It also supports some of the country's highest breeding populations of wood ducks and prothonotary warblers. The refuge serves as home to more than two hundred other species of birds forty-seven species of mammal, and nearly one hundred species of reptile and amphibian. In case that's not enough, Caddo Lake is also considered "the only large naturally formed lake in Texas" and is the ongoing site of an ambitious American paddlefish restoration effort.[9]

Considering this hit parade of ecological features, it's hard to imagine that Caddo Lake could have become anything *but* a national wildlife refuge once the army closed its Longhorn ammunition plant. This, in fact, is a theme common to a number of military-to-wildlife transitions: against all odds, *nature made this happen*. This is an appealing storyline, and the idea that nature bats last—in an affirming way—actually plays a part in many traditional approaches to ecological restoration more generally. Cast this way, the attitude is that in order to bring a degraded or damaged ecosystem back to health, restoration efforts mostly need to give ecological processes and ecosystem components a crucial kick start, then nature will take it from there. This approach comes to view in accounts that explain Rocky Mountain Arsenal's transition from a horrifically contaminated chemical weapons manufacturing plant to a revitalized shortgrass prairie ecosystem that now provides habitat to bald eagles, American bison, and black-footed ferrets. As one FWS publication put it, "In a way, it was the eagles that made it happen."[10] Again, by these terms, scientists, politicians, the DOD, financial calculations, lawsuits, community leaders, FWS officials, and conservation advocates in the end are merely bit players in the larger stage of what is fundamentally a *natural occurrence*.

At Caddo Lake, it's easy to embrace similar explanations that the ecological qualities here were so phenomenal that the only possible destination for postmilitary management had to be conservation oriented. The DOD is perfectly willing to accept and promote this logic; it invites a related view that only by virtue of the military's keen environmental stewardship were these environmental conditions made possible. This, in turn, highlights a broader message that carries real political value—namely, that military activities and environmental conservation are deeply compatible efforts. I wasn't terribly surprised, then, when I met with refuge officials at Caddo Lake, and they all pointed to the important role that the Eagles had played in securing refuge designation out of this former weapons plant. It just wasn't the type of eagle I'd expected to hear about.

When the Longhorn ammunition plant closed, many local business leaders' first response was to push for the site to be redeveloped as an industrial park. There was, after all, adequate infrastructure in place, and the initial loss of jobs from the plant's closing made a new center of local employment sound like a good idea. Perhaps more important, the land came with valuable water rights at nearby Big Cypress Bayou, and an estimated $200 million worth of timber stood on the property.[11] The federal government also held mineral rights for most of the Longhorn site, and natural gas development appeared to be a viable prospect. In other words, there was ample incentive to redevelop the Longhorn site for commercial gain or some form of resource extraction, and plenty of interest in doing so. The idea of turning the former weapons plant into a wildlife refuge, meanwhile, had just a handful of proponents, but one of these was named Don Henley, vocalist and drummer for the Eagles.

As one of the most successful rock bands of the 1970s and '80s, the Eagles may remain best known for their hit song "Hotel California," but Don Henley was born in East Texas and grew up just thirty miles from Caddo Lake, where he fondly recalls catching his first fish. Henley may have found he could never really leave East Texas, and in 1992 he partnered with an attorney friend, Dwight Shellman, to create the Caddo Lake Institute. Ever since, the institute has been dedicated to "protecting the ecological, cultural and economic integrity of Caddo Lake, its associated wetlands and watershed."[12] In just a year, Shellman and Henley had succeeded in securing designation of Caddo Lake as a Ramsar site, identifying it as a wetland of international significance (at the time, it was only the thirteenth Ramsar designation in the United States). Shellman and Henley then proceeded to work their many connections—which included regional wildlife managers, members of Congress, and DOD officials—to press for wildlife designation at the aging Longhorn ammunition plant. The shift at first seemed a long shot, but local residents and federal officials gradually came on board. As one resident and refuge volunteer pointed out regarding the initial effort to turn Longhorn into a private industrial site: "Don Henley and Dwight Shellman had a lot to do with that not happening." Nature, in a sense, played a part; there were, and still are, plenty of reasons to justify protecting Caddo Lake as a refuge for migratory birds, black bear, alligators, and otters, as well as the wetlands and forests that support the wildlife. But as Henley himself later noted, "Sound science may be our saving grace, but oftentimes in Washington—and certainly in Texas—politics trumps science."[13]

My visit to Caddo Lake brought these relationships between politics and the environment more clearly into view, but as I walked the new refuge trails (which were really just the leftover roads from the Longhorn plant), I couldn't help but marvel at the multiple ways this place seemed to intersect with my own life. I'd never even been to East Texas before, but various threads of my past seemed to cross through this place. When I came upon the impassive blast wall where the Pershing missiles expended their final dose of rocket fuel before the army crushed their motors into scrap, I recalled the US airfield behind the village where I lived in Germany one term during college. The locals all referred to the site simply as the "Raketenbasis," and it was widely thought to house Pershings. The plutonium triggers in the Pershings' nuclear warheads had almost surely been made at the Rocky Flats plant, just ten miles from my hometown. Upon the warheads' dismantling, the plutonium would come home again to Colorado, while the motors ended up here in East Texas. Key points in my life's travels, somehow, seemed patterned after those of the Pershing missile and its most deadly components.

It's tempting to think of these overlapping geographies as mere coincidence, or as testament to my unique peregrinations, but the better explanation is the ubiquity of militarization. Most all of us live in such a thoroughly militarized world, often without even noticing it, that we can't help but have many strands of connection that make militarized landscapes very personal. One day as I was working on this book, my local paper ran a story that the water supply for eighty thousand nearby residents is contaminated with perfluorinated chemicals that entered the groundwater from nearby military bases.[14] Located ten miles down-gradient from these installations, many of these households surely never imagined they'd find themselves part of a contaminated militarized landscape that only recently has become evident.

The reach of active military installations can sometimes prove to be surprisingly well hidden, but it is often even more difficult to track the influence of military sites that have closed and subsequently been given new purposes and identities. The meanings and associations for these transitioning sites can change rapidly, sometimes in ways that inspire (think of contaminated brownfields being cleaned and restored to provide wildlife habitat) but also in ways that can obscure hazards, and important lessons and histories. Recasting military sites toward new goals of environmental conservation offers an affirming trajectory for federal lands, often captured in phrases such as "From Swords to Swards" or "Bombs to Birds" that appear as sensible win-win outcomes for transitioning military lands.[15] The military is able to unload obsolete properties, brownfields become new wildlife refuges, and local communities gain open space amenities in place of formerly restricted military bases. The new conservation opportunities that emerge can be quite real but also often remain highly constrained, leaving a number of new wildlife managers ill equipped to deal with military residues and, ultimately, wondering if their military inheritance is more curse than gift. In the next chapter, I turn more directly to the many challenges—and the appeal—of the opportunistic conservation that military-to-wildlife conversions offer.

Most M2W sites require some form of ecological restoration or remediation in order to advance their transitions from military installations to wildlife conservation. Ecological restoration has long struggled with questions about how to properly identify historic baseline or reference conditions, and these challenges become even greater when we recognize the significance of multiple historical and ecological layers of a site. Cultural interests to avoid historical erasure in these cases can work against the aims of ecological restoration, which often seeks to remove prior ecological damage. The prospect of erasure in turn raises questions of authenticity, or how to avoid what some critics call "faking nature."[16] In chapter 3, I consider how one way of working through these problems at M2W sites is to distinguish between the meanings we create when we encounter these places versus the material characteristics and processes of restoration work.[17] The importance of meaning and interpretation thus becomes elevated as the multiple layers and blended (or hybrid) qualities of these places press us to consider more deeply our notions of landscape.

Given these concerns, it also becomes important to examine the degree to which M2W conversions create or disrupt transparency at transitioning military sites, depending in part on the extent of cleanup, new representations of these lands, and new wildlife management or conservation objectives. As I examine in chapter 4, military base closures in the United States provide a chance to open up sites that for decades have functioned as restricted spaces of knowledge and activity, but it remains unclear how fully these sites will emerge as public space in their new designations as wildlife refuges, and how questions of risk will be resolved.

In chapter 5, I turn to some of the many military-to-wildlife transitions occurring internationally, including dozens of national parks and more than three thousand protected areas that have been designated within twenty-five kilometers of the former Iron Curtain death strip that once divided Europe.[18] Much more so than at sites of M2W transition in the United States, European efforts to protect these formerly militarized zones also explicitly attend to preserving the memory of separation and violence imposed on this region during the Cold War. This more integrated approach of cultural and ecological protection provides a useful comparison, with examples of how different kinds of texts, physical remains, and artistic representations can work to keep history "present"; it also highlights the power of landscapes naturalizing in ways that potentially undermine these efforts.

The idea of *military environmentalism* or *ecological militarization* has become one of the central ideas of military-environment scholarship in recent years. This idea emerges from a view that military production and environmental protection are compatible and, more directly, that military activities create conditions to logically transition these sites to ecological preserves. While geographers such as David Harvey have critiqued a related, overarching concept of ecological modernization,[19] in chapter 6 I specifically question the implications of embracing ecological militarization as a model for land management, social change, environmental ethics, or conservation. The casting of militarism as compatible with environmentalism can provide multiple benefits to military interests, conservation groups, local economies, and elected officials, but also carries the risk of greenwashing military activities and avoiding costly, necessary cleanups of contaminated sites. A merger of militarism and environmentalism also jeopardizes cultural preservation, in some cases, and can undermine efforts at social reform and achieving a more democratic environmental politics.

In the concluding chapter, then, I look at how militarized landscapes can be remembered, managed, and restored in ways that embrace environmental gains while also expanding our understanding of these places and the impacts they have borne. Taken together, these ideas highlight how complex characteristics of M2W lands call for site-specific knowledge, transparency in decision-making, and recognizing the legitimate opportunities as well as the very real risks of converting military sites into new, conservation-oriented uses.

My approach in this book can be seen as both case based and conceptual. By bringing a number of particular sites into view, I aim to illustrate the breadth of contexts from which military-to-wildlife transitions emerge, as well as particular challenges and opportunities found in these places. Conceptually, however, it is also important to reach beyond specific examples to understand how these kinds of changes affect not just particular places, but also broader views and understanding. How we come to understand the role of authenticity in ecological restoration, for example, will affect how restoration goals are designed and whether the work that occurs genuinely contributes to progress in repairing damaged or degraded ecosystems. Similarly, our acceptance or recognition of risk in these transitioning militarized landscapes can create lasting effects on the degree to which military activities and broader processes of militarization subject us to long-term hazards that remain unaccounted for, even as surface conditions, land uses, or reputations of militarized landscapes change. If the processes and impacts of militarization are successfully reframed around more ecological outcomes, then the extent of that militarization itself may fade from view and be taken increasingly as simply a form of environmental protection.

Often when I visit these places of military-to-wildlife conversion, I am struck by how "natural" they now appear, though it is also important to recall how they came to exist as they do and to consider them as places that retain, in a sense, qualities of both redemption and sacrifice. The bombs and guns may have fallen quiet in these places for now, but they still echo in many ways across the landscapes they created.

TWO

Bunkers, Bats, and Base Closures

On September 6, 1990, US Secretary of Defense Dick Cheney called together an unusual meeting of environmentalists, military leaders, and other federal officials to discuss the relationship between national defense and the environment. The US Department of Defense generates more pollution than any other institution in the nation, but Secretary Cheney encouraged those he had assembled to focus on a bracing notion of compatibility: "Defense and the environment is not an either/or proposition. To choose between them is impossible in this real world of serious defense threats and genuine environmental concerns. The real choice is whether we are going to build a new environmental ethic into the daily business of defense."[1]

Cheney's remarks signaled several key themes. First, he explicitly rejected making an either/or choice between military defense and a clean environment. Instead, he promoted a vision of military-environment compatibility that would require policy makers to reconsider long-held assumptions that pit human interests against the environment. The field of military geography, for example, has often dedicated itself to projects such as terrain analysis, which assumes militaries must surmount environmental challenges, or use and modify natural features for tactical advantage.[2] Cheney suggested that not only is the business of defense ready for an environmental ethic, but that such an ethic can accommodate progressive military and environmental concerns. With this there is a surprising optimism that protecting the environment and protecting national security will readily find common ground. More fundamentally, it was noteworthy that a secretary of defense was taking the time to concern himself with the military's productive relationship with the environment, and to bring that relationship to the attention of an elite group of policy makers. Cheney's framing of the challenges facing the military and the environment was no doubt intended to convey a particular sensibility to an attentive audience. Ultimately it would become clear that Cheney's keenest interest was to streamline military effectiveness and efficiency, not to become a champion of environmental protection; but the role of his remarks and the ideas these remarks convey prove important as they reach into contemporary connections and changes involving the US military and the environment.

Defense Lands and Wildlife Conservation

In the classes I teach, I often have students who have served in the military. When I ask these veterans about wildlife conservation on defense lands, invariably they have a story to tell. Soldiers who trained at Fort Bragg, North Carolina, recall prescribed burns and the beneficial role of wildfire in maintaining habitat for the rare red-cockaded woodpecker. At the now-decommissioned Fort Ord, on Monterey Bay in California, another student worked on ecological restoration efforts to conserve habitat for the endangered Smith's blue butterfly. Coming out of Florida's Eglin Air Force Base, air force veterans talk about conservation of the threatened loggerhead and endangered Atlantic green sea turtles. Others who were stationed at the army's Yuma Proving Ground and the marines' Twentynine Palms (the Marine Corps Air Ground Combat Center) point to measures to assist threatened desert tortoises.

In recent years, the DOD has spent approximately $3.5 billion to $4 billion annually on its environmental program. In 2016 this included $1.1 billion for environmental restoration and nearly $390 million on natural and cultural resources conservation programs (the remaining funds go to compliance, pollution prevention, and environmental technology).[3] On the nearly 30 million acres it manages, the DOD hosts at least three times the density of endangered and imperiled species compared to that found on other federal lands such as national parks, national forests, or national wildlife refuges.[4]

These numbers offer a credible basis for the idea of military environmentalism: the claim that military activities can serve to benefit environmental quality and ecological conditions. It's certainly true that, much as Secretary Cheney once claimed, we don't always need to choose between environmental protection and defense activities; but the relationship between these two efforts is more complex than the notion of simple compatibility would suggest. For example, some of my students also recount the hassle of needing to deal with wildlife concerns during their military training, remembering concessions to the federal Endangered Species Act not as a win-win for wildlife and the military, but merely as a burden they encountered that disrupted their usual activities. Others scoffed at the idea that environmental concerns could intrude upon the military mission they had been dedicated to, recalling commanders who dismissed environmental regulations and carried on with a sole focus on training exercises or maneuvers.

These accounts perhaps make clear what may already be obvious: conservation of wildlife or habitat resources is not the principal focus of the military, but instead an ancillary consideration. This, of course, is no secret. Even in the US Army's 2014 sustainability report—which largely dedicates itself to touting the new environmental ethic and responsibilities of this branch of the military—a "mission first" credo is evident in the running head throughout the document: "Sustain the Mission, Secure the Future."[5]

To find a federal agency with a *primary* mission focused on wildlife and habitat conservation, we need to look elsewhere. The US Fish and Wildlife Service is one of the least known of any federal land management agency, though its National Wildlife Refuge System spreads across some 850 million acres on more than 560 individual refuges.[6] For most Americans, there is a wildlife refuge within an hour's drive of where we live, whether we realize it or not. In addition to managing the wildlife refuge system, the FWS also carries the major responsibility for implementation of the Endangered Species Act. In 2014, the FWS operated on a budget of $170 million for its endangered species work and $475 million to manage the refuge system.[7] The two federal agencies in focus here—the DOD and the FWS—attend to very different needs through very different means, but there are a number of sites where their interests also very much coincide: the variety of aging military installations in the United States that in recent decades have closed and been repurposed to wildlife and habitat conservation, then were renamed and added to the National Wildlife Refuge System.

BRAC and the Creation of Military-to-Wildlife Refuges

In 1988, the United States convened its first Defense Base Realignment and Closure (BRAC) Commission, which established a new approach to evaluating and closing US military bases that were no longer needed. With four additional rounds of BRAC Commission–directed closures between 1991 and 2005, the DOD has closed or reclassified more than four hundred domestic military bases as result of this process, including more than 140 major installations.[8] Closed military bases are converted to a variety of new uses, ranging from playgrounds or recreational facilities to housing developments, hospitals, business parks, and university campuses; but many

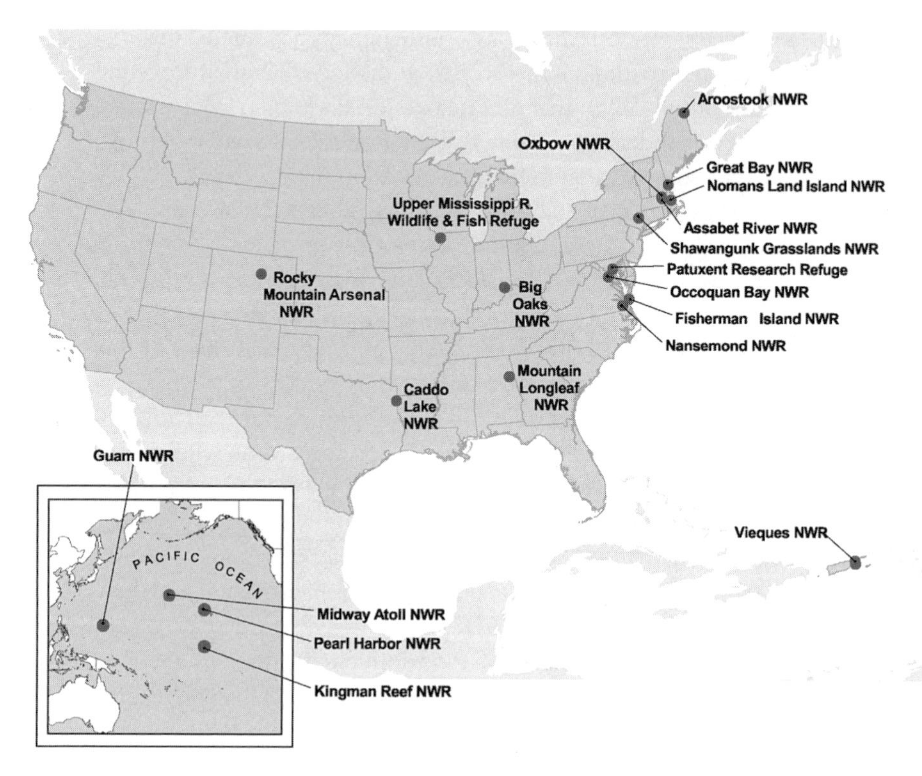

2.1. National Wildlife Refuge conversions from military base closures

military lands face limited options for future use because of chemical hazards, munitions, buildings, or aging infrastructure that remain on-site. Due in part to these reasons, more than 15 percent of the major US bases closed since 1988 have been redesignated as national wildlife refuges. During this period, not counting expansive new marine refuges in the Pacific Ocean, twenty-one bases on more than 1.1 million acres have been transferred to the FWS for management as part of the refuge system (fig. 2.1).[9] With the 2005 round of closures, more BRAC rounds under consideration for the near future, and cleanup or legal issues that often delay closure and conversion, additional military-to-wildlife conversions are sure to take place.[10] Fort Ord, for example, was recommended for closure by the 1991 BRAC Commission, but its long-term status was not resolved until 2012, when President Barack Obama designated the site a national monument.[11]

Of all the options available when a military base closes, rebirth as a national wildlife refuge is not often the most obvious. To convert a former air force base into a commercial airfield, for example, represents a relatively modest change, as does the renovation of army housing to serve civilian residential needs. Such conversions require careful planning and a number of economic, social, and physical adjustments, but the categories of use are in most respects very similar. There is something rather dissonant, however, about turning a bombing range or chemical weapons plant into a refuge for conserving wildlife.

Reclassifications of military lands to wildlife refuge designation in many ways highlight an environmental paradox: DOD lands on the whole are not just more biologically diverse than other federal lands such as national parks or national forests—they are also more contaminated.[12] The coupling of military activities with environmental conservation comes as a surprise to any who view the former as centering on violence and the destruction of life, the latter as dedicated to wildlife and habitat protection. While the primary purposes of the DOD and FWS remain very different, in recent years a new relationship between these agencies has grown increasingly evident. In this view, as Dick Cheney tried to emphasize in 1990, military production and environmental protection should be seen not as oppositional efforts, but compatible ones.

Others joined Cheney in expressing this view. At the July 2000 dedication of the Big Oaks NWR in Indiana, FWS director Jamie Rappaport Clark offered the following remarks to welcome discarded army lands into her agency's system: "Back in the late '80s, I served as the fish and wildlife administrator for the Department of the Army, a job that required me, among other things, to look at how military training exercises could be made wildlife-friendly. It was not as difficult a job as you might think, and so it is not surprising to me that right here on this former military range, we've got an amazing array of wildlife."[13]

Clark's comments, much like Cheney's a decade earlier, suggest that top officials at both the DOD and FWS tend to cast the relationship between defense and the environment in a certain light. There could be many reasons for this, including the simple point that military land restrictions do seem to generate certain kinds of floral and faunal flourishing; but one outcome of this rhetorical "greening" of the military is to obscure some of the damaging effects of military activities.[14] Given this cloaking tendency, it becomes increasingly important to evaluate such official pronouncements of compatibility by asking: How easy is the coexistence between military activity and habitat production? How apt is this depiction, and whose interests does it serve to encourage it?

Opportunistic Conservation

In her critical examination of militarized landscapes, British geographer Rachel Woodward effectively interrogates how militarized sites function as political and ethical landscapes: "The study of military geographies involves a moral decision. If we study the ways in which military activities inscribe themselves onto space, place, environment and landscape, should we ignore or accept unquestioned the politics of that process?"[15]

As the involvement of leaders such as President Obama and Vice President Cheney attests, military land use changes clearly catch the attention of politicians at the highest levels. The particular conservation implications of recent military base conversions and realignments in the United States remain less clear, however, as many of these sites offer conservation and ecological restoration opportunities, but with environmental gains tempered by buildings or hazards that remain from earlier military activities. These factors in many cases limit and shape the conservation efforts directed by wildlife personnel. Even as wildlife refuge officials work to take advantage of the conservation opportunities provided by former military lands, their efforts to open refuges to public use and restore the sites' ecological processes face obstacles. Without structural and institutional changes to address these shortcomings, an M2W refuge may be characterized by conservation gains that are shaped at least as much by the site's military past as they are by the current mission to conserve and protect plants, wildlife, and habitat.

Military Base Closure and Conversion

The DOD prioritizes base closures to improve military readiness and streamline the use of defense funds, but environmental considerations also feature in the process of evaluating and converting sites. Before creating the BRAC process in the late 1980s, the US Congress passed legislation that actually made base closures more difficult.[16] Despite a view promoted by the DOD and others that military lands hold important ecological features, some members of Congress were concerned about the severe contamination of these lands. In effect, they claimed that military bases were simply impacted beyond remediation and too hazardous to convert to other, more public uses.[17] Politics played a cynical part in some of these arguments—elected officials were sometimes loath to see military bases close in their districts, as this often led to short-term job losses and political backlash—but the environmental problems singled out were often very real.[18]

In part as a response to these concerns, legislation in the 1990s required an "Environmental Restoration Program" that identified key sites for preservation or restoration on military lands.[19] This requirement roughly coincided with closure and remediation efforts on Department of Energy lands that had been integral to nuclear production throughout the Cold War.[20] According to US Environmental Protection Agency (EPA) projections at the time, the cleanup of military lands could require the largest environmental remediation ever undertaken by the US government.[21]

A very different portrait of the environmental character of military lands has also emerged in recent years as groups and individuals advocate military conversions *for* conservation purposes. This process often begins only after bases have been identified for closure; increasingly, though, even active bases attract notice from environmental groups, boosters of local communities, or the military itself to highlight the environmental amenities of DOD installations.[22] Dating back to 1962, the DOD has sponsored an annual Secretary of Defense Environmental Awards competition, which honors "innovative work protecting the environment while sustaining mission readiness."[23] Conservation groups, policy makers, and local officials eager for economic development often support reclassifying obsolete military sites into new, more attractive future states as national wildlife refuges, monuments, parks, or other types of open space.[24]

National wildlife refuges are managed with an express primary purpose of conserving wildlife and plants. Unlike most other US federal land categories—such as national parks and national forests, which since the early twentieth century have been created exclusively by legislative action—wildlife refuges may still be established using executive authority (the same is true of national monuments). More than six times as many national wildlife refuges have been established by presidential decree or administrative actions that trace directly back to the executive branch than declared by act of Congress.[25] This approach to designating national wildlife refuges brings spontaneity and flexibility into the system, and has vastly expanded the extent and variety of the United States' network of public lands during the past century, but sometimes comes at the expense of consistency, stability, or purpose. By some critical accounts, the spontaneity of the system at times verges into a state of constantly reacting to external factors, political maneuvering, and the demands of prior land uses and conditions rather than measured planning.[26]

Nearly all refuges in the National Wildlife Refuge System also operate under significant budgetary constraints, and many—whether they are M2W sites or not—find themselves needing to respond to issues caused by prior uses. A 1989 US Government Accounting Office study determined that secondary or non-wildlife-related uses occurred "on virtually every refuge and include all manner of public, economic, and military activities."[27] On refuge lands created out of former military sites, these issues are often complicated by the fact that they not only pose wildlife-related problems for managers, but also create public safety concerns. The challenges of addressing concerns related to prior conditions and public use are only exacerbated by the FWS's chronic shortage of funds. As an incredulous congressional study pointed out in 2011, the agency estimated its refuge system maintenance and operations backlog at $3.3 billion, which dwarfed that year's refuge system budget of $486 million (the ratio has improved somewhat since then).[28]

The pressure placed on FWS land managers to maintain a premier conservation network grows more daunting when lands that need major restoration or careful control of the visiting public come into their custody. Both of these are serious concerns with new M2W refuge additions. These additions can place a strain on the refuge system overall, and also displace funding from existing refuges to pay for more aggressive cleanup activities at prominent, highly contaminated new refuge additions. A former official at Rocky Mountain Arsenal NWR described this asymmetry of federal expectations, appropriations, and site-specific disbursements when I interviewed him in 2004:

> Congress, and the administration now, is telling the [FWS] agency to "curb your appetite for land, we don't have enough money to manage what you have" . . . and then on the other hand they're turning around and giving us land that we didn't ask for that's going to cost more to manage. But no money with it. . . . So, FY '97, '98, '99, just about every new dollar—and those were good years for the refuge budget—just about every new dollar that came into Region 6 came to the Arsenal.

The ongoing process of transferring lands from the DOD to the FWS may appear simply as the continuation of a centuries-old pattern—the redesignation of former military lands into national wildlife refuges has a long history in the United States, dating back to the early twentieth century.[29] But in fact, the pace of transition has accelerated with the recent decades of base closure. Additionally, contemporary M2W lands often come with very different qualities than other national wildlife refuges. For the land managers tasked with reorienting sites from military uses to wildlife and habitat conservation, the challenges and opportunities can seem both unique and quite pressing. These factors limit and shape in important ways the conservation efforts that subsequently take place at M2W refuges.

From Conflict to Conservation

Each M2W refuge is singular in terms of the array of military activities that occurred there, the suite of impacts these created, the ecological features that exist, the remediation or restoration treatments prescribed, and the variety of social interests such as recreational activities, economic incentives, or political pressures that factor into land management decisions. In this way, the cultural and physical geography of the Rocky Mountain Arsenal NWR in Colorado—a former chemical weapons plant in the arid high plains—is dramatically different from the Big Oaks NWR in southern Indiana, which was used as a bombing range and sits in a humid karst landscape of deciduous forest.

Obviously, these and other M2W sites include substantially different qualities that refuge personnel must contend with on a daily basis. However, the broader policy framework in which refuge managers operate is largely standardized. According to the 1997 National Wildlife Refuge System Improvement Act, the system's mission is to maintain a set of lands dedicated to the conservation and (where appropriate) restoration of fish, wildlife, and plants.[30] This mission provides the FWS with the clearest conservation mandate of any US land management agency,[31] and has helped affirm a culture within the agency that reflects a nearly singular focus on fish and wildlife. This shows up in the kinds of programs offered at many national wildlife refuges, adds to the tensions caused by incompatible secondary uses, and is reflected in the natural sciences background common to many FWS employees. In addition to serving the overarching mission of the National Wildlife Refuge System, refuge managers must also direct their efforts toward their own refuge-specific charters, which typically emphasize conservation and habitat concerns, often in the interest of a particular array of species. For example, a former army ammunition storage site in Massachusetts was redesignated the Assabet River NWR due to its "particular value in carrying out the national migratory bird management program."[32]

While these conservation-focused directives for refuges come as a well-meaning (and important) effort to steer management effectively, at M2W refuges with complex histories and confounding habitat conditions that include chemical contamination and leftover munitions, this guidance can also limit what happens at refuges in at least two important ways: first, managers find little institutional support for programs that are not perceived to be "wildlife first"; and second, considering the background, training, and culture of FWS employees, dealing with the remains of military uses rarely qualifies as the most urgent priority. Measures such as the cleanup of military ordnance and other legacies of prior land use are set aside for concerns that support the wildlife and habitat conservation mission more directly. As a M2W refuge official in Virginia commented to me, "If you don't have enough money for your primary job, then how are you going to pay for dealing with [military] bunkers?" Personnel often apply a policy of managerial triage stemming from inadequate funds and wildlife-oriented programs that displace remediation projects.

Active projects at Big Oaks, for example, include prescribed burning to maintain grassland and shrub landscapes, biological studies of the relatively little-known crawfish frog, and annual bird counts. These are each worthwhile, relevant efforts, given the refuge's declared purpose "to preserve, conserve, and restore biodiversity and biological integrity for the benefit of present and future generations of Americans."[33] To their credit, refuge personnel accomplish this work across some 50,000 acres on a shoestring budget. And yet, as I noted earlier, the site remains cluttered by millions of rounds of unexploded ordnance and projectiles, and continues to host a laser-guided bombing range for the Indiana Air National Guard near the heart of the refuge (the bombing range is not considered part of the refuge, but is fully enclosed by it). Elsewhere, a former firing range within the refuge still harbors tens of thousands of pounds of depleted uranium.[34] Technically, remediation of military residues within the refuge remains the responsibility of the DOD, but with day-to-day management turned over to the FWS, there is little incentive for the DOD to act and no dedicated budget for this cleanup. As one FWS manager commented when I met with him, "The army in some ways thought they could just walk away . . . even though the ordnance and depleted uranium and some of these other things are here." Scouring the site of all munitions would almost surely cause more ecological harm than leaving ordnance in place—some munitions are buried dozens of feet below ground—but the DOD has never conducted a thorough survey of munitions remaining even on the surface of the refuge.

Big Oaks personnel readily acknowledge the challenges and limitations of operating a wildlife refuge on the site of a former proving ground, but they typically work to maximize the habitat and wildlife conservation benefits of their actions rather than press the DOD to fully remediate the site. In conducting prescribed burns, for example, one Big Oaks worker explained to me how he needed to change his approach because of pervasive hazards from UXO that limit access to the interior of burn parcels:

> I'm more concerned about the UXO [than the depleted uranium] and yeah, everything's changed—all the standard operating procedure, because of the presence of UXOs, changed. You can't do aerial ignition . . . [you have to do ground ignition] from the perimeter. You can't get in there. Normally you'd be walking straight, like strips across the whole unit. But we have to just move around the perimeter. So we're actually going to try to use a remote control helicopter.

In this way and many others, FWS officials work opportunistically to accomplish tasks that fit their mission. At Big Oaks, the grasslands and openings caused by years of testing military explosives actually represent valuable habitats in their own right, and these now support populations of rare songbirds and amphibians, the endangered Indiana bat, and mammals including bobcat and river otter. Opportunistic conservation in this case provides genuine protection for biological diversity, even if the particular ecological communities and structures present are the product of military activity rather than evolutionary processes operating largely outside of human influence.

This work takes different form in each M2W refuge, depending on the particular restoration activities and the ecological and military characteristics managers face; but opportunistic conservation is similar across sites in the way it represents an effort to turn the material challenges of dealing with military infrastructures or hazards into meaningful measures to protect plants, wildlife, and habitat. This often involves a set of social extensions as well, as managers try to open M2W refuges for public access and a constellation of wildlife-dependent uses such as hunting, fishing, photography, bird watching, and environmental education that the FWS deems important.[35]

Ecological Restoration and Opportunistic Conservation

Although M2W conversions are often pitched as win-win opportunities, where the military becomes more efficient by trimming obsolete holdings and the FWS adds to its conservation land base, the DOD is not particularly concerned with what decommissioned installations *become* after they leave military control. As long as the military offloads the managerial and financial obligations of its unneeded lands, it deems the transaction a success. This view explains one of the reasons that wildlife refuge designations have become a popular response to military closures: the cleanup requirement is typically significantly less to convert military lands to a new purpose of wildlife conservation than if these same lands were repurposed for commercial or residential use.[36] For a wildlife refuge conversion, the DOD must simply meet a cleanup standard that provides for the safety of a refuge worker's estimated forty-hour workweek, not the more continuous exposure of a resident of the site or a commercial property. This makes sense from the perspective of the DOD, but leaves FWS managers to deal with lands that may or may not be well suited to fit into the National Wildlife Refuge System.

Of all recent M2W conversions, the most dramatic case of contamination and cleanup comes from the Rocky Mountain Arsenal NWR, just north of Denver, Colorado. Once reputed to hold the most toxic square mile of land on the planet, the refuge has been the focus of a multibillion-dollar groundwater and soil cleanup funded by the army and a corporate lessee of the site, Shell Oil Corporation (formerly operating as Shell Chemical Corporation).[37] As a result of two decades of remediation and restoration, the arsenal in some ways now merits the "Pearl of the Prairie" label an early congressional sponsor of refuge designation tried to attach to the site (figs. 2.2 and 2.3).[38] Yet even as American bison, mule deer, burrowing owls, prairie dogs, and other native denizens of the shortgrass prairie site now thrive, portions of the land remain too hazardous to ever be released from army control. Conservation is clearly a successful outcome of the legal tangles and years of preparation for military conflict that characterize this site, but even here the actions of managers and the degree of public use continue to be confined to a certain array of opportunities. FWS personnel at Rocky Mountain Arsenal point out that the historical activities conducted on-site create a mix of material conditions and less tangible perceptions they must contend with as visitors continue to associate the site with its military past.

In April 2011, one FWS employee described to me the operational challenges he faced at the Rocky Mountain Arsenal NWR as follows:

> We have a lot fewer [constraints] now, in the sense that the cleanup is really over, other than some groundwater cleanup. And they're [the US Army is] just maintaining their little landfills and so forth. But we used to have lots of constraints from all of the cleanup actions. And we lost a number of areas that were doing really well, from a restoration standpoint, because they decided at a later date, "You know, our soil samples show that right in your restoration area is the best clay that we can borrow and put on the caps-and-covers [landfills]." And they have to do it. We knew that. But we didn't know that at the time we put in the restoration site. No one knew that. We do all this in conjunction with the [US Army and Shell] so there are as few of those kinds of problems as possible. But we still had quite a few. So that is a managerial constraint.

2.2. Rocky Mountain Arsenal chemical weapons manufacturing facilities, ca. 1970s. (Photo by Clayton B. Fraser, Fraserdesign. Courtesy of Library of Congress, Prints and Photographs Division, Washington, DC.)

2.3. Rocky Mountain Arsenal National Wildlife Refuge, Colorado, 2015

As I described earlier with the efforts to manage fire at Big Oaks, for managers at Rocky Mountain Arsenal and other M2W sites, there are almost invariably cases where ecological restoration goals need to be modified to accommodate public safety or access concerns. One FWS specialist working on prairie restoration described how this played out at the cap-and-cover hazardous waste landfills established near the center of the refuge as part of the army's remediation project:

> We had actually spent quite a while developing a plan where we might have been able to graze the caps-and-covers with bison and do fire up there [for prairie restoration] and do all the things that we do with a normal prairie. And we had a plan almost in the [works]. . . . And we got to a point where the state regulator said, "No, we're not going to do that." So right now, yes, there should be some cooperative management up there. Right now the [Fish and Wildlife] Service doesn't have any real authority to go out there and do anything.

Even where the FWS does have clear authority to manage its lands, the agency can run into problems that stem from past military management. Residents of Commerce City, which abuts Rocky Mountain Arsenal to the west, suffered for years from problems of groundwater contamination seeping from the arsenal's chemical manufacturing facilities. With the contaminated groundwater plume now carefully monitored, and clay barriers and groundwater treatment facilities installed on refuge grounds to clean water before it leaves the site, FWS officials report that Commerce City residents and refuge visitors remain wary:

> Sometimes the perception is worse than the reality. And I think that's probably one of the biggest constraints out here. There are a lot of perceptions of this site that exist that is not [*sic*] necessarily the reality of what's here. . . . A lot of visitors come out here and they either think it's radioactive or it's highly contaminated. And, you know, it was only the central core that was really disturbed. When the army bought this as a base, it was the central core. They wanted the rest of it for security. So there's a lot of this land area that surrounds us that really wasn't affected too badly. But public perception here makes it kind of tough.

Most M2W sites have not seen the level of contamination or cleanup that has been conducted at Rocky Mountain Arsenal. At several refuges in the northeastern United States, the most tangible reminder of prior decades of military management is dozens of hardened concrete bunkers (or "igloos") (fig. 2.4).

These pose little risk of contamination, but even as inert relics they present managers with incongruous markers of militarization within their new conservation landscapes. The DOD has no commitment to maintain or demolish these structures, so FWS officials employ them in a variety of ways: for vehicle storage, as experimental habitats, as historical attractions. Or they simply neglect them.

The lack of DOD involvement itself draws a diverse set of responses from FWS managers: many would like more from the DOD—to see more cleanup and remediation work, and more reliable funding—but a few indicate that they would prefer *less* DOD activity, essentially to support opportunistic conservation. In a 2011 interview, one refuge manager explained to me how some in the FWS view the US Navy cleanup of the Nomans Land Island NWR as a potential threat to wildlife and plants that inhabit the site. The refuge, a small island off the coast of Massachusetts, is off-limits to all public use since it remains contaminated by UXO after being used for aerial gunnery from 1942 to 1996.[39] The navy has a continuing obligation to remove UXO from the island, which brings remediation and conservation objectives potentially into conflict with each other and raises the question whether additional disturbance as part of remediation or restoration efforts is worth the ecological harm, at least in the short term, that it can generate.

In some cases, military hazards can be used as a type of management tool. A refuge official at a M2W site in Virginia told me: "From an enforcement and public education standpoint, it's much easier to keep people out of refuges when there's a public safety concern, such as UXO, than for biological reasons." At other M2W sites, the military residuals may actually pose more of a public *attraction* than hazard. At Assabet River, for example, the fifty concrete bunkers formerly used by the army for ammunition storage have become a significant lure for visitors, with volunteers offering bunker tours. These military artifacts can be more of a draw than the migratory birds and waterfowl to which the refuge is dedicated. As one Assabet River official acknowledged: "I would consider [the bunkers] an attraction. . . . They're more popular than anything else here, more so than the birds."

Even where cleanup is viewed as essential for public safety or reasons of environmental contamination, such restoration efforts also create impacts that affect conservation—at least in the near term. At the wildlife refuge created from the former navy gunnery range on the island of Vieques, Puerto Rico, for example, clearing UXO even superficially from the 14,000-acre eastern portion of the refuge requires removal of the top foot of soil and clear-cutting all trees smaller than three inches in diameter.[40] These treatments significantly reduce vegetation cover, change habitat conditions, and

2.4. An ammunition igloo remaining at Assabet River National Wildlife Refuge, Massachusetts

increase soil temperatures and erosion, but may allow for expanded public use at least on roadways in the interior of the refuge (public use is currently limited to a handful of beaches on the perimeter). The aesthetic impacts rendered by these surface treatments are also substantial, and the full extent of public use on the refuge remains in doubt, as deeply buried munitions still litter the site (fig. 2.5).

At Vieques, years of military training and testing create an uneasy relationship with conservation efforts. The same is true at other M2W refuges. Long term, the militarization of these places is the reason other land uses have not developed. While the transformative effect of military use is sometimes not very apparent—when groundwater or soils remain contaminated, or when munitions lie buried deep in the ground—it may be just as lasting as more obvious land use changes such as agriculture or commercial and residential development. And in a very real sense, the militarization of these lands is fundamentally responsible for the mixture of qualities, negative and positive, that now make them available for inclusion in the National Wildlife Refuge System. As a manager at New Hampshire's Great Bay NWR (formerly Pease Air Force Base) observed: "If the military hadn't come in in the '50s, this would probably look nothing like it does today. In a sense, the military presence has made this what it is—the fact that it is a large, unbroken tract of land with six miles of undeveloped shoreline. It wouldn't have happened. It would have been nothing like this."

A number of wildlife refuge officials recognize the transformative effects of prior military uses of these lands. At times there can be synergistic effects between military activities and wildlife management. One FWS employee at the Great Bay NWR recalled work the DOD completed: "The fish ladders in this water control structure, they were put in by an engineering unit from the air force as a training exercise. I think we probably paid—we paid for the structures and materials, but they did all the labor, and the equipment they provided. It's that thinking outside the box, that creative solution" which makes this place work.

There are, in fact, credible examples demonstrating that fires lit to clear munitions testing areas (or ignited inadvertently by the ordnance itself) can create useful habitat openings and grasslands, that buffer zones associated with militarized sites provide valuable open space habitat, and even that tank tracks, bomb craters, and concrete bunkers can provide microhabitats for wildlife.[41]

2.5. Surface clearing for unexploded ordnance at Vieques National Wildlife Refuge, Puerto Rico

But if viewed uncritically, these examples of military-environment compatibility can be treated too generously, used by the military or other boosters to simply overwrite an array of adverse impacts caused by militarization.[42]

Regardless of whether refuge managers are inclined to be accepting or critical of military environmentalism, those officials tasked with directing conservation and restoration projects at M2W sites do face real challenges. As I explore more fully in the next chapter, even fundamental questions such as determining historic reference conditions for restoration goals become fraught with difficulty at these sites with complex histories.[43] As a refuge manager at one former bombing range acknowledged when I met with him, "We're certainly interested in what was here, and what the opportunities are from what was here. But because of the soil depletion and changes and the surrounding landscape and the exotics, we have limitations. We can't just forget about those things. Sometimes those things are more important than what was here in the past."

Speaking more pointedly about the predicaments he faced when dealing with military contamination on a refuge created from a former air force base, a now-retired refuge manager in the northeastern United States pointed to systemic problems: "The DC office, the director [of the FWS], they really haven't put money in contaminants. To me, I think half of this [wildlife biology] stuff is just ridiculous. I mean, look at the kind of work we're doing here. If you haven't done a contaminant analysis on these species, I don't think you should be doing anything. That's my bottom line." This tension between conservation aspirations and the lingering military presence leaves some managers wondering where their management objectives fit in the landscape and what they can or cannot do to work for wildlife. One M2W manager reflected when I asked for his take on M2W transfers, "Did we really want this [land] in the beginning? I don't know."

Opportunistic Conservation: Challenges and Possibilities

As the phrase indicates, opportunistic conservation makes use of *opportunities* available at a given site—but is also characterized by significant constraints. Military lands can limit wildlife managers' ability to plan systematically and to implement management policies without the disruption or surprise of encountering new hazards or sites of contamination. FWS officials at M2W refuges periodically find themselves reacting to unexpected problems: munitions brought to the surface by rain or erosion, wildlife burrowing into contaminated areas, or otherwise routine remediation work uncovering hazards abandoned by earlier military operators. To cite just one dramatic case, in October 2000, contractors at the Rocky Mountain Arsenal NWR were cleaning a pile of scrap metal when they uncovered a grapefruit-sized bomblet filled with lethal sarin gas. The site was quickly closed off for more careful excavation, and the next month, nearly a dozen additional sarin bomblets were discovered.[44] The nerve gas was eventually destroyed without further incident, but the episode brought the disturbing hazards of the new refuge back into public view. There have also been subsequent disruptions to refuge operations—including the 2007 discovery of the blistering agent lewisite at a cleanup location, and in 2013 an extensive closure due to flooding when a small dam burst after record-setting rainfall.[45]

At times such as this, managers need to react quickly and may find themselves veering from the restoration work or conservation planning they anticipated. In these cases, opportunistic conservation may seem a burden and a departure from what the Society for Ecological Restoration in the past has cast as ecological restoration's focus: "intentional activity that initiates or accelerates the recovery of an ecosystem."[46] Although the organization in some cases still points to a traditional set of goals for restoration that often privilege historical conditions, the Society for Ecological Restoration also acknowledges that contingencies can emerge: "Restoration attempts to return an ecosystem to its historic trajectory. Historic conditions are therefore the ideal starting point for restoration design. The restored ecosystem will not necessarily recover its former state, since contemporary constraints and conditions may cause it to develop along an altered trajectory."[47]

This emphasis on intentionally restoring a site to be historically faithful to prior conditions runs into obstacles in a wide array of settings, and has proved to be a constant question in ecological restoration: *which* prior condition should restoration efforts target? In complex landscapes with multiple layers of human and natural history, or areas with particular hazards (such as M2W refuges), entire categories of restoration activity may be unavailable. Restoration planning at these sites, instead, will often tilt toward the reactive. Managers can act intentionally or thoughtfully in their broad planning and even with most daily tasks, but the physical conditions at complex sites such as former military installations often limit how reliably they can focus on restoring ecological integrity. Such limitations are not only found at former military sites, to be sure; but the unique hazards and infrastructure created by proving grounds, missile bases, weapons manufacturing, or ammunition depots create a suite of challenges rarely encountered by most restoration ecologists.[48]

It seems worth emphasizing that these factors point to a broader concern about how complex landscapes work at the interface of conflict and conservation. At these sites, by virtue of the ongoing need to deal with military hazards, the military-environment relationship can remain biased toward the military even long after the end of military activities. Outwardly, M2W refuges change dramatically as the names and management goals reflect new FWS control, but the refuge managers' actual operations must still attend—at times as their most immediate guiding interest—to the legacy of militarization. A material presence of the military very much remains in the form of chemical contaminants, munitions, massive concrete structures, or other military infrastructure, but these remains are often obscured from public view both by the site's explicit representation as a space dedicated to wildlife and conservation, and by restricting public use to areas that are free of these components. As a result, against the sanguine view of an emerging military ethic of environmental awareness,[49] a number of sites that would seem to most embody this convergence of militarization and conservation actually may in important ways serve to misrepresent how the military affects the environment.

Sites such as the Rocky Mountain Arsenal, Big Oaks, or Vieques are now acclaimed for their habitat amenities, many of which were either created or protected by military activity. However, these places also retain military hazards essentially in perpetuity, creating a mixed legacy of contamination and conservation. Even at former military sites such as Great Bay or Assabet River that today seem relatively lightly impacted, managers continue to grapple with what to do with the infrastructure bequeathed by prior military activities. As a manager of a M2W refuge with similar concerns in Maine noted to me, "Our job is to manage for wildlife, not to keep a coat of paint on a concrete bunker."[50]

Acting opportunistically to pursue their conservation mission, M2W refuge managers often work diligently and creatively to create safe opportunities for the public use of these lands, to develop environmental education programs, and to make these sites function compatibly within the broader mission of the National Wildlife Refuge System. In this way, a concrete bunker formerly used to store Nike Hercules missiles may be modified to serve as a bat cave, or 1.2 million cubic yards of contaminated soil and debris can be consolidated and buried at the center of a refuge, capped, and covered with native grasses.[51] While these transformed facilities remain militarized—the double-lined hazardous waste landfill at the Rocky Mountain Arsenal, for example, is one of the few parcels within the refuge boundaries slated to remain forever under army jurisdiction[52]—they also act as spaces of conservation. This, then, is the lingering paradox of M2W refuges: they exhibit some of the most persistent qualities of conflict and militarization while also revealing new possibilities and opportunities.

There surely are risks that come with the conflation of militarization and conservation, including the historical erasure of the processes and policies that produced some of the worst contamination found on the planet. In highlighting the conservation value of former military sites, M2W refuges may make less obvious the possibilities that have been foreclosed for these places. Yet there are also risks to writing off militarized landscapes as unmitigated disasters that must be relegated to permanent status as brownfields. A visit to the site of the former Pease Air Force Base or Rocky Mountain Arsenal or the Longhorn ammunition plant—each now a national wildlife refuge—certainly brings a different experience today than it would have several decades ago. With a conservation mission now foregrounded, military activities are largely absent. Opportunistic conservation can thus emerge as a cynical move to cover the tracks of military negligence or as a genuine and creative effort to achieve conservation successes. In either case, it makes sense to work to understand these places in their complexity, to engage with them as sites of conservation *and* militarization, and to do what we can to ensure that we face the challenges posed by such complex landscapes as comprehensively as possible.

One important set of approaches that can help turn contaminated or heavily impacted militarized landscapes to function as new conservation-oriented sites is the science and practice of ecological restoration. Defined by the Society for Ecological Restoration as "the process of assisting the recovery of an ecosystem that has been degraded, damaged, or destroyed,"[53] ecological restoration works to bring together physical and natural sciences with other forms of knowledge—including those of local or indigenous populations—in ways that can bridge the traditional nature-society divide. By integrating knowledge and practice, ecological restoration efforts have the potential to effectively respond to the many complexities of militarized landscapes.[54] As I consider in the next chapter, however, military-to-wildlife land use changes also challenge in important ways some of the basic ideas of restoration efforts.

THREE

Real Restoration?

Just north of Denver, Colorado, the Rocky Mountain Arsenal NWR offers a respite from the buzz of traffic on Interstate 70, the roar of aircraft from Denver International Airport, and the refineries and railroads that have long marked nearby Commerce City as Denver's industrial hub. Designated in 1992, the wildlife refuge today protects 15,000 acres of shortgrass prairie, wetlands, and riparian corridors in what the US Fish and Wildlife Service suggests is "one of the finest conservation success stories in history."[1]

To the casual visitor looking across an expanse of freshly burnt prairie, or a narrow band of cottonwoods tracing the line of a dry stream channel, this casting of the arsenal as a world-class conservation site might seem a bit overblown. The refuge scarcely fits the North American conservation ideal of wild landscapes, rugged peaks and canyons, or untrammeled nature. The high-rises of Denver loom nearby, fences and roads break the refuge into discrete parcels, and even the lakes scattered across the southern tier of the site are clearly hemmed in by dams and levees—in some cases capturing storm runoff from urban Denver.

Although the refuge sits as an island of habitat in a matrix of urban, industrial, and agricultural activity (the FWS describes the site as one of the nation's largest *urban* wildlife refuges), the site also legitimately brims with wildlife. Bald eagles roost in the cottonwoods; white pelicans and cormorants ply the waters; white-tailed and mule deer roam the refuge, in spring often trailed by coyotes looking for a stray fawn; a growing herd of bison paws at the high-plains grasses; and thousands of prairie dogs plow the soil, contributing in turn to habitat or food for burrowing owls, badgers, hawks, and, since an October 2015 reintroduction effort, black-footed ferrets.

All this makes for a remarkable conservation showcase, perched as the refuge is amid the jumble of North Denver; but what qualifies Rocky

Mountain Arsenal as a truly world-class conservation success is its history. For nearly four decades, starting in 1942, this was the site of a US Army chemical weapons facility. For much of that time, Shell Chemical Corporation and other lessees also produced industrial pesticides and poisons. By the 1970s, the arsenal had earned a lurid reputation not just for manufacturing some of the world's most lethal chemical and nerve agents, but for its poisoning of local communities and groundwater, and triggering the largest earthquakes in Denver's recorded history.

The latter, caused by the army's short-lived program of injecting toxic waste into two-mile deep wells on-site, was how I first heard about the arsenal—news related to me by my fifth-grade teacher, who found it impossible to ignore stories of this site, just thirty miles away from school, that had gained infamy for harboring the most toxic square mile on our planet. When I tour the site today, as I often do with my own students, I am inclined to think of its story not simply as one of conservation success, but as a cautionary tale of Cold War devastation followed by decades of intensive ecological restoration. In other words, it speaks to our often-broken relationship with the environment, as well as our periodic determination to try to make amends.

Prompted by court order, since 1990 the US Army and Shell have completed more than two billion dollars' worth of restoration to transform the arsenal from a site dedicated to military weapons and chemical production to one dedicated to wildlife and habitat conservation. The work has been neither simple nor cheap, but the steps taken here—and at other locations undergoing similar transition from militarization to conservation—bring into view the promise of ecological restoration and some of the many challenges it faces when confronted by complex and historically layered landscapes.

A Traditional View of Ecological Restoration

Ecological restoration brings an array of approaches to M2W refuges as restoration scientists, refuge managers, citizens, and conservation groups work to change the character on the ground of former military sites and reorient these places to new conservation and wildlife habitat goals. The successful management of M2W refuges often depends on some form of military cleanup and ecological restoration. The complex histories of these landscapes pose challenges, however, to traditional views of restoration as a process that seeks to bring back the prior form and function of an ecosystem.

Practices that today are used for ecological restoration date back to early human history—think of indigenous use of fire to restore and maintain prairie or clear forests of underbrush—but the science of restoration ecology is much more recent.[2] The term *restoration ecology* was not coined until the early 1980s, and only toward the end of that decade did the first academic journals emerge, along with a professional organization. From the outset, restoration ecologists struggled to set the terms of their endeavor. In 1990, the nascent Society for Ecological Restoration (SER) defined ecological restoration as "the process of intentionally altering a site to establish a defined, indigenous, historic ecosystem."[3] Three years later, SER modified this definition to something more practicable: "Ecological Restoration is the process of re-establishing to the extent possible the structure, function, and integrity of indigenous ecosystems and the sustaining habitats they provide."[4] During this same period, the National Research Council defined ecological restoration as "the return of an ecosystem to a close approximation of its condition prior to disturbance."[5] Other restoration scholars and practitioners pointed to the *Oxford English Dictionary* for a definition of restoration: "the act of restoring to a former state or position . . . or to an unimpaired or perfect condition."[6]

Each of these definitions rests to some degree on the idea that restoration efforts should point toward an earlier, undisturbed, idealized pristine state. This proves problematic on multiple counts, including critiques that are practical, ecological, and philosophical.

In practical terms, the 1990 definition set an impossibly high bar: Could humans genuinely re-create an indigenous, historic ecosystem? And even if this were possible, wasn't there an impossible contradiction in saying we would dedicate human labor to the task of returning a landscape to a condition unaffected by human activity?[7]

On ecological grounds, restoration definitions that point to a perfect, predisturbance condition no longer fit current understandings of ecosystem dynamics. Rather than presenting ecosystems in static, "climax" conditions, by the 1990s the "new ecology" demonstrated that most ecosystems are actually characterized by flux rather than stasis.[8] As this view gained favor, it became increasingly difficult to argue that restoration ecology should set as its goal some particular historic condition. If ecosystems were bound to change, wouldn't it be arbitrary—and in a sense, *not ecological*—to select a snapshot in time and try to fix those conditions into place? And even if this could be seen as desirable, which time should we choose for that snapshot? In North America, pre-European settlement was often used as a logical historical target, but what to do, then, for restoration efforts *in* Europe? And what were we to make of North Americans' pre-European standard in light of clear evidence that indigenous North Americans substantially modified ecosystems for thousands of years?

Finally, on philosophical grounds, early definitions of restoration seemed problematic for suggesting that human traces on the land (at least, post-European human traces in North America) needed to be eradicated in order to bring back some version of a prehuman world. This seemed to assume that people should necessarily be considered apart from nature, rather than being a part *of* nature, and reduced the role of humanity on earth to one of destruction.[9] Restoration ecologists viewed themselves as trying to reestablish a more positive relationship between humanity and nature,[10] but even this often seemed grounded on a notion of human-nature separation.[11]

By the late 1990s, then, restoration ecologists found themselves challenged on multiple fronts to defend their work at its most basic level: What ought to be the goals of ecological restoration? Clearly, a new definition was needed to reposition restoration objectives, and the new definition would somehow need to take into account a shifting understanding not just of the science of restoration ecology, but also of how we sought to position ourselves more basically with respect to nature. In 2004, SER's Science and Policy Working Group came out with a new definition that cast ecological restoration as a *process* to assist the recovery of ecosystems that have been "degraded, damaged, or destroyed."[12]

Restoration and History

The revised definition of ecological restoration in many ways has been liberating. Without orienting restoration around a fixed, seemingly arbitrary goal to attain a historical reference condition, restoration practitioners can focus more on ecological function than the form, or composition, of an ecosystem. For example, if an area formerly known to be shortgrass prairie has become encroached by shrubs or trees, restoration focusing on "assisting the recovery" of this site can emphasize restoring historic processes, such as fire, to the system rather than worrying about the particular components or concentration of individual species. The restoration work may still be motivated by a vision of turning shrubland to prairie, but ecological processes are given a chance to do what they may to achieve this goal. At Rocky Mountain Arsenal, this approach comes through in the seasonal application of prescribed burns and in using nonnative annual cover crops, such as sorghum, to reduce soil loss, outcompete invasive weeds, and fix atmospheric nitrogen in the soil prior to replanting native species.

The 2004 SER definition of restoration also accommodates social considerations in restoration more readily. By shifting the focus of restoration goals to assisting ecosystem recovery rather than creating a particular assemblage of plant or animal species, it creates openings for human participation in that process. People, in this way, can be brought directly into view as part of restoration efforts, both as restoration participants and in a sense as restoration objects. Turning again to Rocky Mountain Arsenal for an example, managers at the site distinguish between two broad areas: restoration goals for the northern portion of the refuge target a shortgrass prairie that approximates presettlement conditions and a very traditional approach to ecological restoration, but goals for the southern third of the refuge include managing and maintaining an array of features earlier residents of the site put in place. These include a series of lakes that are now used for recreational fishing, waterfowl habitat, and urban storm-water retention; windbreaks of nonnative trees and shrubs planted by nineteenth- and early twentieth-century settlers; and the terminus of the historic High Line Canal that was completed in 1883 to bring water from the South Platte River's Waterton Canyon more than sixty miles to North Denver. With these and other features, restoration goals at Rocky Mountain Arsenal include at least some accommodation for human layers or histories and not only those considered natural or ecological.

This approach, of opening ecological restoration to goals that are not necessarily tethered to a particular condition or time, creates new room for maintaining cultural landscapes or spurring broader conversation of what restoration ought to do, but it also brings into view an important question: What is the appropriate role of history in ecological restoration? When we reorient ecological restoration to goals based on a process that disregards historic reference conditions, do we unhinge the whole effort to the point of restoration itself becoming arbitrary? Scholars who now point to the need to embrace "novel ecosystems"—for example, sites dominantly influenced by changing processes of climate or urban development—argue that historic reference conditions simply may not be relevant or attainable, given intense or broad-scale ecological transformations.[13] Others suggest that restoration ought to be future oriented, perhaps angling to satisfy new combinations of human and ecological interests, so historic conditions need no longer serve as a guide.[14] These approaches make many restoration ecologists uneasy, with some wondering if an ahistoric approach to restoration represents a fundamental enough challenge to the field to merit new labels. Perhaps rewilding, a term gaining favor especially in parts of Europe, captures this sense of change that is not so bound to history; or we could call it remediation, which my own students seem drawn to for its rather vague promise of simply trying to make a damaged site better.

A look at actual ecological restoration efforts in the United States suggests that traditional approaches aiming for a pre-European condition remain the norm. Despite the variable treatments taking place at Rocky Mountain Arsenal, the stated focus of the FWS at this site sounds very much like earlier versions of SER's restoration definition: "The U.S. Fish and Wildlife Service has spent many years, with many more to go, to restore the land to as close to its native condition as possible."[15] Interviews I conducted with the refuge's personnel revealed a similar orientation toward traditional goals of restoration. As one of the key officials in charge of prairie restoration at the site explained, "The overall goal here with restoration is to do the best job we can in turning Rocky Mountain Arsenal back into native prairie." When I asked another high-level refuge official what he would consider "authentic" restoration at the site, he responded, "Bringing back exactly what used to be here before Caucasians showed up."

A similar privileging of pre-European conditions comes through in visitor attitudes at this site. During a four-month period in 2010, with a team of researchers I conducted surveys of visitors to the refuge. When asked which landscape condition would be "most natural" to the Rocky Mountain Arsenal site's history, more than 70 percent of the respondents selected "uninhabited prairie for plants and wildlife." (All other choices, including "recreational facility," "Plains Indians settlements," "farmlands/homesteads," and "chemical manufacturing," each drew 10 percent or less of the responses.)[16] When we asked a related question, "What landscape condition would be 'most faithful' to the Rocky Mountain Arsenal's site history?" more than half of the respondents chose, "uninhabited prairie for plants and wildlife."

This suggests that even as the main professional society dedicated to ecological restoration has repositioned itself for more than a decade around a definition for restoration that relates to history only indirectly (referring to some prior degradation or damage), for land managers and at least this portion of the visiting public, ecological restoration still conjures visions of a frontier unsullied by European colonization.[17] This leads to questions of what really counts as "authentic" restoration, and how historical, human attributes and uses of a landscape can be integrated with ecological goals and restoration targets that seem historically defensible.

Authenticity and Historic Fidelity in Ecological Restoration

The question of authenticity has presented a chronic challenge to ecological restoration, and was brought pointedly into view with a critique by philosopher Robert Elliot in 1982. In his article "Faking Nature," Elliot likens ecological restoration to art forgery: much as the forged painting appears to be something it is not—an original work by a particular artist—ecological restoration presents a place or ecosystem that appears to be shaped by nature but is actually the work of restoration ecologists.[18] Even if the restoration creates an ecosystem that is an exact replica of the site prior to its human impact, Elliot argues this is still a landscape with diminished value. With its natural genealogy disrupted by human action, the restored landscape can never be considered the genuine or authentic article, it remains a "fake" posing as real nature.

Other restoration scholars and practitioners take issue less with this question of originality and unbroken natural lineage, and raise a concern of restoration goals becoming too centered on human interests or desires. In this critique, by stripping away a focus on the historic ecological baseline for a site and turning instead to human interests or a shifting current or future condition, "we are likely to fall prey to the mass consumerism that surrounds us—creating gardens where we maintain beings as 'things' strictly for our use and admiration."[19] Unlike Elliot's view, which contends that restored sites will always be fakes, this second view allows that restoration can be authentic if it genuinely attends to the historical, predisturbance ecology of a given site. The hazard, then, is not that creating an original ecosystem is impossible, but that we would choose the *wrong* condition—either poorly informed by prior ecological conditions or simply veering too much toward human interest and influence.

The ecological restoration of militarized landscapes to new priorities of wildlife conservation necessarily intersects with each of these concerns, but in ways that differ from most other kinds of restoration sites. For a typical prairie restoration project, for example, land use changes often involve a relatively straightforward historical progression: agricultural use (or in some cases, urban development) "broke" the intact prairie, and restoration seeks to reverse that process. The human impact on the prairie—plowing and planting, or other forms of development—is viewed as a form of degradation that needs to be halted and fixed in order for earlier ecological systems to thrive. Ecological restoration in this way orients fundamentally around a project of repair, to rid a site of the disturbances that knocked it away from an original, "natural" condition.

Militarized landscapes and other sites with complex land use histories include additional steps in their historical progression. At Rocky Mountain Arsenal, for example, European settlers displaced native tribes (who, of course, also modified the prairie in important ways, including hunting and the use of fire), then converted short- and mixed-grass prairie into farms and ranches. In the early 1940s, the farm families were in turn displaced by the US Army, which converted agricultural land into an intensive chemical weapons production complex. Four decades later, the army ceased operations and Congress declared the site a national wildlife refuge. Wildlife refuge officials tasked with managing the site by the early 2000s then faced multiple possibilities for ecological restoration: they found general consensus that erasing the impact of the chemical weapons facility should be a priority (in fact, this was mandated as a result of legal action brought by the State of Colorado to address concerns about contaminated groundwater and public health), but what of the previous historical layers? Would it make sense to restore the site to its 1930s pre-*military* condition of agricultural land and farmsteads, or should restoration focus on the more traditional goal of a pre-*settlement* state dating back to the early nineteenth century? Which of these would represent the most authentic form of restoration?

If we fully adopt Elliot's critique about restoration as "faking nature," then all attempts to return the site of Rocky Mountain Arsenal to a previous condition would come up empty. Any restored ecosystem would still exist as mere replica of the naturally evolved prairie that settlers disrupted nearly two centuries ago, and thus would lack the value of the original. Perhaps this is so, but even if we agree with Elliot's argument entirely, the specific details of Rocky Mountain Arsenal—and many similar sites of military or industrial impact—would seem to warrant a different perspective. After all, if remediation efforts and ecological restoration were shrugged aside by these philosophical concerns, one can scarcely imagine arguing that the site would best be left alone, to remain conceptually unsullied—no false representations here!—but ecologically damaged as a perennially toxic brownfield leaking contaminants into the area's groundwater.

As a matter of policy, then, it likely makes sense to remediate and restore the Rocky Mountain Arsenal site even if that means the result will not be as valuable or "authentic" as the original prairie that evolved there.[20] This leaves society still to grapple with the second concern, however: Which prior condition should we use as a target for restoration goals? Most traditional restoration ecologists have a ready response for this second concern: restoration should aim to at least restore key ecological processes so a site can return to something approximating its presettlement conditions. In the case of Rocky Mountain Arsenal, this would point to the shortgrass and mixed-grass prairie utilized for centuries by the indigenous Arapaho, Kiowa, and Shoshone, and populated by wildlife ranging from bison and pronghorn to wolves, ferrets, and prairie dogs. Ecologically, this approach treats two significant phases of landscape disturbance—agricultural production and chemical production—as stages in a broader period of degradation, and seeks to erase both from view.

As I mentioned already, this in fact is the underlying premise for most of the restoration that has occurred at the refuge. Wildlife officials use fire, herbicides, mowing, seeding, and other active management techniques to facilitate the return of prairie vegetation across much of today's wildlife refuge. The FWS has also worked to restore native fauna to this site, including a herd of bison reintroduced in 2007 (and now naturally reproducing) and black-footed ferrets. With many other native species such as prairie dogs, badgers, coyotes, mule and white-tailed deer, bald eagles, ferruginous hawks, and burrowing owls still extant at the site, today's refuge is only missing wolves and pronghorn from its main presettlement palette of wildlife.

Given the dramatic history of this site, however, it seems worth asking how fully traditional restoration objectives ought to guide efforts at Rocky Mountain Arsenal and other sites with complex or culturally significant histories. If we successfully return this place to a condition that appears unchanged from the way it looked when the Stephen Long Expedition encountered it in 1820, however we rate this as an ecological accomplishment or a conservation success story, we might still raise legitimate concerns about the meaning this restoration conveys. At a broader level, this success might carry with it a fraught lesson of technological supremacy, that no matter the insult we bring against the environment, we can remedy the damage and restore a site to a healthy, full condition (Elliot refers to and rejects this as the "restoration thesis"). While this concern may seem minor in cases where the original damage was relatively slight or temporary—a logging road that is later removed, for example, or a field that was cleared and now restored to native prairie—when the human impact is dramatic or has wide-reaching social or ecological implications, then this out-of-sight, out-of-mind approach becomes more problematic. These cases would surely include many former sites of intense industrial or military activity; thus we might want to guide restoration efforts to promote some forms of erasure but work to avoid others.

At Rocky Mountain Arsenal, there is general agreement that environmental remediation and ecological restoration efforts have made this site significantly safer, although some citizen groups have persistently raised serious concerns.[21] Across much of the refuge, cleanup efforts have transformed a dangerously blighted chemical production facility into an increasingly popular open space and conservation amenity. The erasure (or, more accurately, the consolidation and containment) of chemical hazards at Rocky Mountain Arsenal, in other words, has become a widely touted ecological success. However, the cultural implications of this cleanup remain less clear, as a number of local residents and others worry that the site will increasingly become known only as a wildlife refuge and that its prior condition as a chemical weapons site and location of agricultural homesteads will be lost to history. My own research and surveys of visitors supports some of these concerns, which are also registered by FWS managers at the site and local groups such as the Commerce City Historical Society.

Even as officials at Rocky Mountain Arsenal work to manage the refuge as a restored, functioning prairie ecosystem, many of them also acknowledge there is a risk of the arsenal's history being lost from view. The refuge visitor services staff I interviewed explained their approach as follows:

> We're telling [visitors] the whole story. We're trying to communicate to them everything that's gone on here; what we—the Fish and Wildlife Service—as a whole are here to do now. That is challenging, because there are many, many stories to this place on many different levels, so I think it is our job as visitors' services staff to try to give people a better understanding of what happened here, and it's a lesson that's learned. There's a lot of lessons learned so that these kind of things don't get repeated in the future. And "These are the type of actions that have affected the land, and this is the result of those actions" I think it is our duty to tell people that. We're not trying to brush it under the rug.

There are, in fact, a number of exhibits in the Rocky Mountain Arsenal NWR visitor center that vividly portray the history of chemical production at the site, as well as displays that document the periods of Native American use, agricultural settlement, and subsequent displacements that preceded military and industrial use. A significant percentage of visitors never make it into the visitor center, however—a self-guided driving tour of a portion of the refuge opened to the public in 2012, which allows visitors to bypass interpretive exhibits. The visitor center is also open just Wednesday through Sunday from 9 a.m. to 4 p.m., even though the refuge itself is open daily, dawn to dusk year-round.

A number of residents of nearby Commerce City, immediately west of the refuge, and Denver's Montbello neighborhood on the refuge's south border remember the site during its active chemical production phase and consider this history to be of critical importance. When I met with members of the Commerce City Historical Society and asked what they thought today's refuge visitors should learn about the site, there was widespread agreement that the full story of what happened here should be told:

[FIRST RESPONDENT]: First, you'd know the history of it from the beginning all the way [to now].

[SECOND RESPONDENT]: I agree. . . . Not only [did the US Army] displace those people [farmers], but on top of that, what they brought were terrible problems to our area—and we had to fight so hard to get it fixed. We're very thrilled that it's a reserve and all cleaned up and all of that. But we would love them to keep the history alive out there. I'm a person that thinks that the history makes it what it is. . . . That's just a real story to tell how the people here were so strong, that they came to do that [settle the land]. Because if they hadn't done that, this [wildlife refuge] would not be here today.

Others from the group expressed dismay that virtually all physical remains from the army's decades of occupation and production had been removed from the site. One example was fresh in the minds of group members when I met with them, as the FWS had recently removed the security gates that had long guarded entrances to the refuge (refuge officials were concerned that the gates could be intimidating to nearby residents and scare off potential refuge visitors). The historical society group's broader sentiment was one of wanting to keep the multiple histories of the site tangibly in view. As my conversation with the group unfolded, it became clear that these local residents appreciated more than just the nature of this site. They offered a list of features that might also have been usefully restored or preserved, including buildings, cemeteries, and schools:

[FIRST RESPONDENT]: . . . So that when they had their tours they could say, "This is where Rose Hill Cemetery was, and this is where the school was."

[SECOND RESPONDENT]: I think it would've been cool to leave a water tank.

[THIRD RESPONDENT]: There could be farming equipment sitting out there, and old buildings.

[FOURTH RESPONDENT]: I think the gates should've stayed up, or even [been] moved somewhere along the [road].

If we look beyond the concerns of these longtime neighbors of the site—a number of whom also recall adverse effects of foul odors and poisoned wells from the chemical production that took place—there are also national and international implications that call for more than prairie ecosystems being restored or maintained here. Many of my students in recent years have been shocked to learn that for decades the United States assiduously developed an array of chemical weapons that we now find abhorrent. Sarin gas, for example, which was manufactured and stored at the Rocky Mountain Arsenal, recently made international headlines when evidence indicated that Syrian President Bashar al-Assad had used it against his own people. Surely there are lessons we could all learn from this period of militarization, as well as the decisions that ultimately led to the (ongoing) dismantling of our chemical stocks and forswearing their production and use.

We might also learn valuable lessons from this site simply in thinking through the easy connections that long existed between chemical weapons development for military application and commercial chemical production for home and industrial use. As Wendell Berry once asked in a 1989 commencement address, "How would you describe the difference between modern war and modern industry—between, say, bombing and strip mining, or between chemical warfare and chemical manufacturing? The difference seems to be only that in war the victimization of humans is directly intentional and in industry it is 'accepted' as a 'trade-off.' Were the catastrophes of Love Canal, Chernobyl, and the Exxon Valdez episodes of war or of peace?"[22]

In other words, there is a great deal we can, and no doubt *should*, learn from sites such as Rocky Mountain Arsenal that we put at risk if we dedicate our efforts here exclusively to ecological restoration goals. This comes into view even more forcefully at sites where the previous layers of use remain closer at hand.

Vieques, Puerto Rico

Vieques NWR spans nearly 18,000 acres of the small island of Vieques, located eight miles east of the main island of Puerto Rico. The refuge includes slightly more than half the area of the entire island, which is also home to about ten thousand people. Prior to the first wildlife refuge designation in 2001, the US Navy used Vieques for six decades of artillery testing and munitions storage. Beginning in 1941, the United States cleared thousands of the island's residents off the western and eastern portions of the island—some with as little as ten days' notice—in order to make room for military activities. Unlike with Rocky Mountain Arsenal or similar domestic cases of military land expropriation in advance of US involvement in the Second World War, at Vieques many of the displaced residents remained nearby and persisted in their claims that when the navy was finished, residents should be allowed to return to their land.[23]

Instead, when the navy halted its activity on Vieques in 2003, all its remaining land was turned over to the FWS to be managed as a wildlife refuge (fig. 3.1). In 2005, portions of Vieques Island were added to the US National Priorities List as one of the nation's most hazardous "Superfund" sites. By December 2015, the US Environmental Protection Agency reported that 2,500 acres of the former navy artillery ranges had been cleared of more

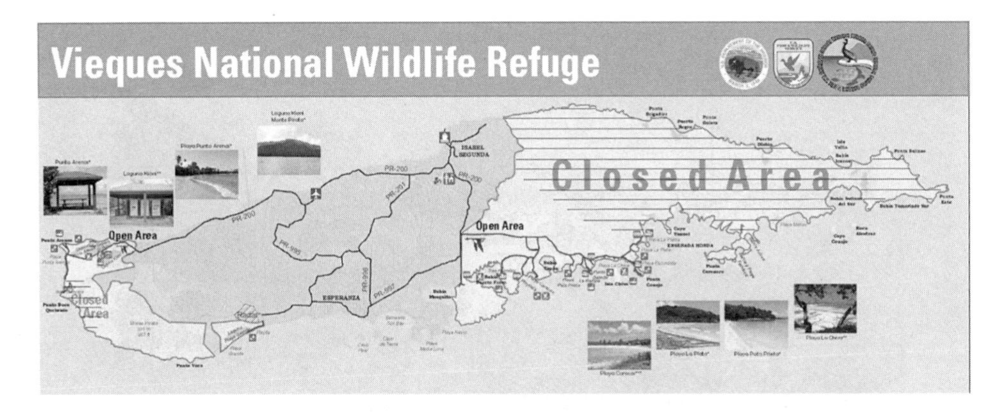

3.1. Map of Vieques National Wildlife Refuge, Puerto Rico

than thirty-eight thousand munition items in an ongoing cleanup and restoration effort that is expected to run through 2025 (fig. 3.2).[24] Considering that the US Navy estimates that its training exercises fired more than three hundred thousand "munition items" on Vieques from the 1940s to 2003, the cleanup thus far would leave nearly 90 percent of all munitions still in place or unaccounted for.[25]

The cleanup operations at Vieques have been problematic in a number of ways. Costs continue to escalate, from a 2003 US Navy estimate of $30 million, to $130 million in 2005, to $350 million in 2012.[26] Some critics of the cleanup estimate that costs will run as high as $450 *billion* to accomplish a full cleanup.[27] Finding and removing unexploded ordnance is time-consuming and hazardous, and made more difficult in beach and shallow marine environments that continually expose (or hide) ordnance over time. On land, UXO removal requires widespread clearing of vegetation, creating impacts on soils, plants, animals, water flow, and aesthetics. The navy has argued that in some cases, the ecological impacts of UXO removal outweigh the benefits, and in the eastern portion of Vieques where the navy operated its live-fire range, cleanup is targeting only the top foot of soil and top four feet of beach sands. Considering how UXO can be liberated by heavy rain and storm events, even after "cleanup" many of the currently restricted areas of the island will remain off-limits to people. As anthropologist Katherine McCaffrey points out in her critical analysis of Vieques, "The most devastated terrain, the 980-acre live impact area, is officially designated as a 'wilderness preserve' and blocked from public access. . . . Land designated for 'conservation use' requires only a superficial cleanup, since presumably no humans will inhabit it."[28]

3.2. Spent munitions casings, Vieques National Wildlife Refuge, Puerto Rico

This highlights one of the key concerns of military-to-wildlife land conversions: that they may simply serve as a least-cost option for the Department of Defense to dispose of unneeded lands while at the same time providing an impression of sound environmental stewardship. In 2009, the DOD awarded the Vieques cleanup team its Secretary of Defense Environmental Award for Environmental Restoration, citing innovative UXO identification methods, an accelerated cleanup schedule, and the successful return of a few select lands to the municipality of Vieques (a number of these sites were transferred after the navy determined no cleanup action was required).[29]

Even as the FWS works to implement new conservation-oriented objectives at Vieques and the US Navy continues its cleanup operations, the longtime residents of the island continue to express resentment over the federal government's expropriation of their lands in the 1940s, what they see as broken promises to allow residents to return once navy artillery operations ended, and in some cases what appear to be wildlife managers' even more restrictive regulations (fig. 3.3).[30] Looking at the wildlife refuge, local people now find themselves prohibited from subsistence uses of fish, crabs, and other wildlife that were sometimes allowed during the period of military activity, and they often turn their resentment against the new federal managers of the site. It's hard to overstate the challenges facing FWS employees: they must make the best of complex situation with ecological and contamination requirements at a site that came as an unrequested "gift" from Congress.

Vieques has made national headlines not only for widespread protests calling for the navy to stop its shelling of the island, but also out of concern that the navy's actions have poisoned a number of the island's residents. In 2009, the US agency responsible for assessing health hazards at the nation's most contaminated sites reversed earlier statements that years of military activity at Vieques had caused no health risks. Citing missing pieces in important scientific data about contamination on the island, the director of the Agency for Toxic Substances and Disease Registry (ATSDR) noted, "The gaps we have found indicate that we cannot state unequivocally that no health hazards exist in Vieques."[31] In fact, a follow-up report issued by the agency acknowledged that Viequenses suffered from elevated levels of

3.3. "FWS NO" sign, Vieques National Wildlife Refuge, Puerto Rico

chronic diseases (including asthma, heart disease, and various cancers), but then cited limitations in the analyses of the Vieques population. The agency concluded that "new and previous data still could not identify a relationship between military activities and health problems experienced by the island's residents."[32]

Independent scientists have criticized the government reports on a number of counts, and highlight how studies can be designed in ways that make it virtually certain that they will fail to find certain kinds of results.[33] In 2010 testimony to a US House Committee on Science and Technology, John Wargo, a Yale professor of risk analysis and environmental policy, noted, "A careful review of the ATSDR public health assessments reveals an agency determined to find no causal relation between the Defense Department's sixty year history of dropping nearly 100 million pounds of weapons on a small island, and the exceptional incidence of human illness among those that lived through this history."[34] Wargo and other researchers have described multiple pathways that heavy metals and other contaminants in munitions can follow to expose residents to toxins, including air, water, soil, and terrestrial and marine foods. UXO found in the former live-impact zone of Vieques is routinely detonated as a means of disposal by the navy cleanup teams, so even the ongoing cleanup efforts may be contributing to island residents' exposure to military contaminants. In 2009, the mayor of Vieques reported that the majority of emergency room visits on the island were for respiratory problems—an astonishing condition for a small Caribbean island with relatively few cars and no major industries.[35]

Adding another layer of complexity to the political and ecological layers found at Vieques is the fact that tourists and tourist-oriented investors have popularized the island since 2003, drawn by its undeveloped beaches and growth opportunities. As a 2007 *New York Times* article explained, post-2003 "Vieques quickly became the next 'It' island—luring travelers with its newly demilitarized white sand beaches, roaming wild horses and rustic charm. As its popularity has taken off, new restaurants and boutique hotels have opened up, raising the level of luxury on the island along with its prices." In other words, not only have longtime residents failed to get their land back and likely suffered health impacts from the years of artillery practice on this island, but now they face new pressures of gentrification and a growing (mostly mainland US) tourist presence.[36]

Turning again to ideas of restoration and authenticity, the diverse interests found at Vieques point to very different perspectives on what restoration ought to look like here. To the FWS, appropriate restoration centers on conserving native plant communities; wetland habitats; and fish, wildlife, and plants representative of Vieques "prior to major agricultural and military use of the land"[37]—in other words, trying to move the Vieques ecosystem toward a traditional restoration goal that precedes significant human impacts. The FWS's "Comprehensive Conservation Plan" for Vieques also assumes that the refuge lands will be "cleaned of any contaminants that would pose a threat to either the wildlife or visitors,"[38] a condition that appears to be far from imminent, given current navy cleanup plans and funding.

To Viequenses, however, authentic restoration entails a thorough cleanup of chemical and explosive contamination, then accommodating some semblance of returning to agricultural, subsistence, or even residential uses. Ongoing human impacts are very much a part of many Vieques residents' vision for the future of the island's refuge lands, whether that means subsistence agricultural use or hunting and fishing, more intensive grazing or cultivation, or ecotourism and other tourism-related uses. These contrasting and largely conflicting perspectives on what future conditions the restoration of the Vieques NWR ought to achieve pose obvious problems for refuge managers and their relationships with island locals. Ironically, the key area of overlap between Vieques residents and FWS managers—ensuring a comprehensive cleanup of existing military contaminants—is perhaps the one aspect that the navy is least likely to grant. One of the strongest rationales for the congressional handoff of Vieques to the FWS, after all, was the fact that wildlife refuge and wilderness designations would necessitate less-rigorous cleanup standards than a return of agricultural, residential, or many tourism-related designations.

In Vieques, the points of divergence between the navy, the FWS, and local residents are varied, though clear concerns emerge that relate to economics, ecology, and politics. The military seeks to minimize costs and maximize efficiency as it relieves itself of unneeded military lands; wildlife managers are directed by law to focus on ecological objectives; and Viequenses aspire to end a lingering colonial relationship with the federal government and renew uses and revenues from half of their island that has been essentially off-limits for decades. These differing interests may be seen as logical outcomes of the specific contexts of this place, but there are also broader questions of historic erasure and commemoration that apply here—as in other sites where militarization is either ceding to or overlapping with new conservation-related goals. What, for example, should future generations come to understand about sites like Vieques? Are there certain cultural attributes or histories that we should be dedicated to keeping in view? And what are the likely consequences of losing sight of certain histories, of allowing restoration to erase not only ecological harm, but also cultural legacies of these kinds of places?

Erasure and Commemoration

A few minutes before dawn on July 16, 1945, the United States detonated the world's first atomic bomb above the southern New Mexico desert. The location of the atomic bomb test, Trinity Site, today is part of the US Army's White Sands Missile Range, the largest terrestrial military installation in the Western Hemisphere. A variety of weapons testing, missile, and rocket programs continue at this site, but twice a year, one day every April and October, Trinity Site opens to the public, attracting caravans of visitors to this iconic epicenter of the nuclear age. Although it may seem surprising that an active military base opens in this way to accommodate history buffs and military tourists, army officials explain the tradition in terms of its historic importance: "As stewards of this significant historical site we want to ensure the public continues to have access to Trinity Site so that we can continue to share this story with future generations."[39] As historian Ryan Edgington has documented, the White Sands Missile Range also includes an array of features that have made it an important conservation landscape, but these ecological attributes remain overshadowed by the militarization of the site.[40]

This balance between militarization and conservation tilts differently at many of the places I have visited, sites where new conservation objectives can steadily erode the public understanding of former military activity. At Trinity Site, the flash of the world's inaugural nuclear test remains bright enough to keep its history in view, but what of places such as Vieques or other M2W refuges, which year by year develop reputations (and infrastructure) to support their naturalization as places devoted to conservation? What might be lost if we fail to commemorate what happened at these sites and their histories fade?

The answer, of course, varies depending on the context. But a look at a handful of sites that have converted relatively recently from military objectives to conservation goals highlights some of the possible consequences of historical erasure that can come as a result of ecological restoration. At Vieques, the local people view commemoration as a means of the federal government keeping active claim to appropriated land. A number of Viequenses continue to press for a complete military cleanup from the live-fire and other affected areas for public health and safety reasons, as well as the prospect of reinstalling premilitary agricultural and residential uses of off-limits areas. There is some risk in this position, however—it assumes that remediation and ecological restoration will lead to more open management policies and increased public uses of the national wildlife refuge. The FWS's Comprehensive Conservation Plan does, in fact, call for increased public access and use of refuge lands, including regulated harvest of sea grape and coconut, land crabs, marine fish, and doves, pigeon, and waterfowl, but all of these are predicated upon the removal of military ordnance and expanded funding for the refuge to support sufficient staff to manage visitors. With a DOD oriented toward streamlining its castaway assets efficiently, and a chronically underfunded National Wildlife Refuge System whose maintenance backlog routinely runs several times higher than its annual operating budget, both these goals may prove to be elusive.

The refuge's conservation plan also notes that public uses of the refuge will remain subordinate to the primary goal of wildlife conservation: "An overriding concern reflected in this plan is that wildlife conservation is the first priority in refuge management. The Service allows public uses if they are compatible and appropriate with wildlife and habitat conservation. The refuge would emphasize wildlife-dependent public uses (e.g., hunting, fishing, wildlife observation, wildlife photography, and environmental education and interpretation)."[41] Under the most plausible scenario of a partial cleanup, with its goal of clearing beaches of ordnance to a depth of four feet and inland soils more superficially, more beaches might become available for recreational use, but Vieques most likely would remain a wildlife refuge that accommodates limited public access and little to any use of its interior areas beyond roads. If the refuge's beautiful, undeveloped beaches continue to draw tourists in increasing numbers, it's easy to imagine a future where Vieques becomes ever more widely known as a gem of Caribbean beach destinations, and less and less a site of antimilitary and anticolonial protest. What Viequenses may find over time, under this scenario, is that their island becomes normalized as a yet another Caribbean vacation site for mainland tourists, with diminishing concern about locals' rights of return to use or occupy a significant portion of their land.

Military-to-wildlife conversion sites on the US mainland obviously involve a different dynamic than that found on Vieques. Great Bay NWR, in southeastern New Hampshire, seems well removed from the specter of colonialism that lingers over Vieques, but even here the initial creation of Pease Air Force Base and its later conversion to a wildlife refuge raise important questions of commemoration and erasure. Like many domestic US military installations, Pease has its own story of displacing locals and disrupting preexisting communities. In this case, the call to assert eminent domain and

3.4. Church parsonage property, Newington, New Hampshire, abutting Pease Air Force Base and its barricaded road

commit individuals' and town property to create a military base came in 1952 during the strategic buildup to the Cold War. The town of Newington, New Hampshire, was abruptly appropriated, and to this day portions of the town center and its main roads can be found in isolated remnants scattered across the current wildlife refuge and airfield (fig. 3.4).

Pease Air Force Base was quickly mobilized as a Strategic Air Command base, serving as home to a fleet of long-range bombers prepared for Cold War strikes. Only after the base was closed in 1991 (a casualty of an early BRAC Commission recommendation) did many residents of nearby Durham and Portsmouth, New Hampshire, learn that Pease had served as a storage depot for nuclear weapons. As such, it was almost surely one of the Soviet Union's priority targets in the advent of military conflict, a prospect that led many in the area to reconsider how broad goals of national security scaled down to *insecurity* at the local level.

Another factor that shaded nearby residents' views of Pease Air Force Base was its legacy of contamination—in this case, derived not from chemical weapons or explosives, but from the more mundane military hazards of jet fuel, degreasers, and solvents that were used and carelessly treated during air force maintenance operations. In 1990, the Pease site was added to the National Priorities (Superfund) List, and monitoring and cleanup efforts have been under way ever since. In June 2014, the state of New Hampshire and the federal EPA documented continued contamination of groundwater in the aquifer underlying Pease (where the airfield is now operated as a commercial trade port and New Hampshire Air National Guard base) and directed the air force to address a range of persisting environmental problems it had caused.[42]

Even as the site is very much at the center of these broad concerns over historic preservation and ongoing risks posed to local residents, a visit to the Great Bay NWR can feel like a trip back to nature. Footpaths lead visitors to Upper Peverly Pond and the shores of Great Bay, where the refuge protects "one of the longest stretches of undeveloped shoreline" found in the region.[43] In a variety of ways, the site clearly fulfills its pledge to "provide wildlife and visitors alike with a place to rest and recharge."[44] It remains less clear how these salutary attributes fit with the legacy of displacement, risk, and contamination. One jarring but perhaps appropriate representation of the mixed character of the Great Bay NWR comes at its center, where the visitor parking lot, refuge offices, and trailheads abut the impressive remains of the air force's high-security enclosure. Inside the fences, a dozen reinforced concrete bunkers that once housed nuclear missiles now offer storage for FWS equipment, provide experimental habitat for bats, and offer hiding cover amid patchy shrubs to support a growing population of New England cottontails. The military-era water tower looming above the enclosure stands pockmarked with drill holes of woodpeckers that have hammered through the exterior insulation looking for insects (fig. 3.5).

These remnants left behind by the air force at Great Bay provide a reminder that this site has not always been a wildlife refuge. In this way, some of the historical uses of the site remain in view; but to many visitors these artifacts also lend themselves to a triumphant account of nature's perseverance and the compatibility of military management with conservation objectives. Though there is truth to these accounts, the darker legacy of Newington's dislocation and ongoing underground contamination goes largely unseen and uninterpreted.

The idea that cultural attributes of a landscape are worth commemorating is scarcely novel. The US National Park Service (NPS) is best known as an agency dedicated to conserving the spectacular scenic landscapes of Yellowstone, Yosemite, the Smoky Mountains, or the Grand Canyon, but this sibling agency of the FWS also finds room in its mission to protect historic sites and cultural memories important to the United States. At the Gettysburg Battlefield, for example, the Park Service "has committed itself

3.5. Enclosure and trail, Great Bay National Wildlife Refuge, New Hampshire

to preserving and caring for Gettysburg National Military Park as a symbol of America's struggle to survive as a nation and as a lasting memorial to the armies and soldiers who served in that great conflict. Our job is not only to preserve the battlefield but to provide you, the park visitor, with a fulfilling and memorable experience."[45] Some NPS sites commemorate events or places that hardly qualify as highlights of a triumphant American history. Little Bighorn Battlefield National Monument, where General Custer made (and lost) his last stand against a coalition of Native American forces, is charged with providing visitors "an understanding of the historic events leading up to the battle, the encounter itself, and the consequences by both the military and the American Indian contingents."[46] Regarding the Trail of Tears National Historic Trail, meanwhile, NPS asks visitors to "remember and commemorate the survival of the Cherokee people, forcefully removed from their homelands in Georgia, Alabama, and Tennessee to live in Indian Territory, now Oklahoma."[47]

NPS historic sites and battlefields typically dedicate themselves overtly to cultural remembrance and commemoration, but as protected areas often found in urban or otherwise heavily modified landscapes, these sites can also conserve important environmental attributes. Ecologists Todd Lookingbill and Peter Smallwood call these ancillary conservation measures the "collateral ecological values" of parks. They note that over time, many of these sites set aside primarily for their cultural and historical significance have also provided a variety of ecosystem services, even though these services were not initially identified as part of land managers' mission or cited as reasons for the land designations.[48]

Much like the earlier example of Trinity Site, at NPS-managed historic parks, trails, and battlefields the focus is on the cultural significance of these places; but there is also room to value the ecological amenities that often accompany the agency's protective measures. In many cases, the NPS intentionally attends to both cultural and ecological components—they are viewed as mutually supportive efforts to conserve landscapes that are both cultural and, often in a restorative sense, natural. The Park Service can do this in part because its enabling legislation, dating back to 1916, articulates a vision of preservation. This includes natural and historical components: "to conserve the scenery and the natural and historic objects and the wild life therein and to provide for the enjoyment of the same in such manner and by such means as will leave them unimpaired for the enjoyment of future generations."[49] Beyond this legal underpinning, however, NPS management of natural and cultural attributes depends on a view that these are not entirely distinct domains, but rather ought to be considered more holistically, as components of complex or hybrid landscapes.

This notion of social and natural interconnection is a bit more elusive in the FWS, which continues to take pride—justifiably in many ways—in managing "the only system of federal lands devoted specifically to wildlife."[50] The National Wildlife Refuge System mission, which was not established until 1997, emphasizes this dedication to wildlife, habitat, and restoration, and the professional culture of the agency also supports this mission. In fact, the traditional route to find work with the FWS has been to earn a degree in fish and game management or wildlife biology (the NPS, by contrast, first stocked itself with landscape architects).[51]

Given the policy directives and professional traditions of the FWS, it makes perfect sense that many refuge managers point to their "wildlife first" mandate as a reason not to manage for cultural features of their refuges. As one manager told me when I asked about the importance of managing sites of military history found on his refuge, "Someone's special shrine—for someone who lived or worked at this site [during its period of military use]—to us may be just a falling-down building that's going to cost us a lot of money."

These views make sense and, especially considering the severe budget limitations refuge officials face year after year, enable managers to focus on their core priorities. There is, however, an important political effect of this approach, at least in refuges where important social histories risk being lost from view. If all we come to know about refuges is that they now provide important protections for wildlife, then we will have lost important insights about how we have prioritized national defense, the sacrifices this required of people and the land, and (in some cases, at least) the callous disregard of these sacrifices the military demonstrated. We need not approach former military sites in a disjointed manner, where we either raise them up as national monuments and historic sites or remove decades of military impact from view by declaring them new places of nature. In very real ways, these sites remain both militarized *and* natural (or naturalizing), and they may help us better realize the perils of isolating these as separate domains.

Even within the terms currently in place for managing national wildlife refuges and the FWS's "wildlife first" creed, there is room for a broader approach. Among the guiding principles Congress set for the refuge system in 1997, it identified "wildlife-dependent uses involving hunting, fishing, wildlife observation, photography, *interpretation, and education*, when compatible, [as] legitimate and appropriate uses of the Refuge System" (my emphasis).[52] These so-called secondary uses of refuges surely could be leveraged as part of restoration efforts to the benefit of cultural commemoration and to offset the risks of erasure at refuges where important social histories warrant it.

To get a sense of how this might work for military-to-wildlife refuges in the United States, we need only to look at the work of Michael Cramer, the member of the European Union Parliament who was instrumental in establishing the Iron Curtain Trail through Central Europe. When asked about ongoing efforts to turn these formerly militarized borderlands into the Green Belt of Europe, Cramer emphasized that we mustn't consider such places in terms of nature alone, but rather need to account for ecological changes in the context of the culture, politics, and history of these places.[53]

In addition to concerns about the loss of cultural memory and the erasure of human history that ecological restoration of militarized sites can present, these land use changes also pose a different kind of hazard: the prospect that systematic, institutionalized risks will become naturalized in ways both tangible and theoretical. In the next chapter, I turn to related questions of how these newly defined conservation lands function as public spaces, and how they serve as examples of the way unanticipated costs—or risks—can be bound into modern industrial societies.

FOUR

Sanctuaries Inviolate

Driving across the expansive Chesapeake Bay Bridge-Tunnel, it's easy to miss Fisherman Island NWR. The refuge covers a 2,000-acre splotch of dunes, wetlands, and stunted trees that marks the north entrance to the bay. When I first tried to visit Fisherman Island and its refuge in 2010, I missed them both entirely. When I realized my mistake, after I'd crossed a final short bridge and arrived on the mainland of the Delmarva Peninsula, I turned around and drove back to Fisherman Island, the southernmost of Virginia's barrier islands. Once there, I discovered nothing but a small sign and a locked gate marking the refuge.

I shouldn't have been surprised. Although national wildlife refuges are the category of federal public lands most likely to exist within five miles of any United States resident, many of these sites do not have a visitor center, public facilities, or full-time paid staff. Public use is typically limited to daylight hours, and must meet a compatibility standard to ensure that the primary mission of the refuges—wildlife and habitat conservation—is upheld.

The Fish and Wildlife Service describes the history of Fisherman Island as a place where "protection is an important theme."[1] Indeed, this theme has appeared in multiple layers during the past dozen or so decades. The island was first used by the United States, beginning in 1868, as a quarantine facility to isolate contagious sailors and immigrants and protect Chesapeake ports from the spread of disease. Later, during the First and Second World Wars, Fisherman Island was fortified with bunkers, heavy artillery, and a network of explosives that extended into Chesapeake Bay to protect against enemy attack and submarines. Most recently, protection at Fisherman Island has focused on wildlife and habitat: Fisherman Island NWR was established in 1969, and in 1973 sole ownership rights were transferred to the FWS by

4.1. UXO sign from Mountain Longleaf National Wildlife Refuge, Alabama

the Department of the Navy (later transfers and acquisitions rounded out the refuge in 1998 and 2000).[2]

If you want to visit the Fisherman Island NWR today, assuming you can find it, you need to sign up for a guided tour, held Saturday mornings from October to March. The tours are free, but on the other six days and six months of the year, the refuge remains off-limits to the public.[3] FWS officials will sometimes acknowledge that former-military-sites-turned-refuges provide a doubly powerful rationale for limiting public use: first, there is the common explanation that human activities and visitors can disturb sensitive species—in the case of Fisherman Island, ground-nesting birds, for example; but second, the public can be waved off on account of military hazards that remain on-site, whether in the form of decrepit buildings, unexploded ordnance, or chemical contamination. As an agency employee at Fisherman Island candidly explained during my tour of the site, this second factor is often the more potent way to secure public compliance with refuge regulations. It's one thing, after all, to inadvertently step on a plover's egg, but even thrill-seekers on their new four-wheeler might think twice about driving over a nest of grenades. A number of military-to-wildlife refuges highlight these risks, both because such risks can be very real and because it can help refuge managers enforce regulations against public use (fig. 4.1).

The motives behind alarming signage are often mixed—some postings seem designed primarily to satisfy liability concerns—but restrictive management prescriptions bring into question the degree to which refuges created out of military sites function as lands that are truly public, at least in the sense of being *open* to the public. Most all national wildlife refuges limit the range or timing of recreational activities in order to protect the public goods of wildlife resources and habitat, but very few refuges beyond those impacted by military activities need to limit public use in order to protect public safety. This raises several important questions. First, if military impacts and hazards persist at these new refuges, how complete is their transition away from a militarized condition? Next, if this transition is only partial, then what is the appropriate treatment of these new refuges? And finally, what do lasting limitations on the uses of military-to-wildlife refuges signify in terms of how the public understands and appreciates these places?

According to legal scholar Robert Fischman, national wildlife refuges sit in the "middle of the permissible uses continuum of the federal public land systems."[4] This means that wildlife refuges, with their assortment of conservation-focused and "wildlife-dependent" secondary uses, fit between the very limited restricted uses of military lands and the relatively wide-open multiple-use lands of national forests and vast western tracts controlled by the Bureau of Land Management. Conditions at some M2W refuges, however, suggest that this middle ground has not really been found. At Fisherman Island, for example, the military-to-wildlife transition and its associated shift to conservation still have little to show thus far in terms of movement toward the "public use" side of the spectrum. In places such as Vieques, the shift from military use to wildlife refuge has even resulted—at least for now—in *increased* restrictions on public use.

Given the conservation mandate of the National Wildlife Refuge System, this careful regulation can be extremely helpful in fostering conservation efforts and discouraging illegal use. However, keeping new wildlife refuges largely off-limits to the public also clashes with the assumption that as military installations are closed and redesignated to other purposes, this will lead to an increased openness to public access and greater understanding of how these lands are being managed. This mismatch of expectations and subsequent regulation can also raise suspicion or resentment by local populations who may have pushed for military closure or refuge designation, but are then surprised to find that lands remain or become largely out of reach.

If the public is to trust federal land management agencies and support their activities, they will likely insist on a legitimate understanding of why particular regulations exist and seeing some semblance of their expectations being met.[5]

For local populations awaiting new recreational opportunities or the opening of former military lands, the transition in management between agencies can lead to confusion or frustration. In Madison, Indiana, more than a decade after the establishment of nearby Big Oaks NWR, locals still commonly refer to the site by its old name: the Jefferson Proving Ground (or JPG). In some ways this makes sense. For most locals, their relationship to this place has little changed since the army transferred daily control to the FWS. The site is closed to the public entirely from December through March, and during its eight-month public use season is open only every Monday, every Friday, and the second and fourth Saturdays of these months. Visitors are required to take an annual safety briefing focused on the presence of UXO across the refuge and to sign a liability waiver; they are then permitted to use just a small portion of the refuge surrounding Old Timbers Lake. The interior of the refuge is open only more broadly during turkey- and deer-hunting seasons each fall. Despite FWS officials' aspirations to make the refuge more accessible to the public (with a simplified entry process and visitor center), the Department of Defense has no plans to remove ordnance from the refuge, and FWS funding priorities have yet to focus on expanding opportunities at Big Oaks.

Concerns about risk remain central to the management of almost all former military installations as they transition to new uses. The decision to convert a military site to conservation purposes is often fundamentally based in calculations that determine that the cost of remediating and restoring these lands is prohibitive if they are to be truly opened to public use and access. The estimated $2.1 billion dedicated to opening the Rocky Mountain Arsenal NWR in even a limited way to the public is indicative of the commitment required. Estimates of what a full cleanup would cost at Big Oaks easily run into the billions of dollars as well. The military's 1995 disposal and reuse analysis for the site estimated that removing millions of rounds of UXO would cost from $8,500 to $22,000 per acre for a clearing depth of four feet. To remove UXO to a ten-foot depth (which would still leave some ordnance in place), the estimates jump to $30,000 per acre for a "best case" scenario, up to $88,000 per acre for less optimal conditions.[6] At this rate, cleaning the entire 50,000-acre refuge could cost upward of $7 billion in 2016 dollars. Little wonder, then, that this and a number of other M2W refuges remain either largely or entirely closed to public use.

Managing for Wildlife, Limiting Use

Some form of public use is permitted on nearly every national wildlife refuge,[7] but public access has never been an established right or even a priority interest of the refuge system. The first US wildlife refuges were specifically designated to protect wildlife *from* the public; for example, at Pelican Island, Florida, where the rampant hunting of egrets, herons, pelicans, and other "plume" birds was threatening local populations with extirpation. Initially this refuge and others were necessarily kept off-limits to public use simply to ensure the survival of the species in question—in some cases, it required armed vigilance to enforce these protections.[8] By 1929, when the Migratory Bird Conservation Act was passed, Congress described refuges as "inviolate sanctuaries for migratory birds or as refuges for wildlife."[9]

Although the US wildlife refuge system has gradually been opened to more and more uses—including hunting on a majority of its units—at many M2W sites the conservation mission compares favorably to that of most other refuges due to restrictions on public use. At locations such as Nomans Land Island NWR off the coast of Massachusetts, or units in the Pacific Remote Islands refuges, public use is prohibited completely—a management approach that some consider the best possible result for plant or wildlife conservation. One refuge manager of a converted air force base described this approach as the military's ultimate version of wildlife conservation: simply put a fence around the land and call it protected. Whether or not a hands-off approach truly is the most effective for conservation remains a matter of some debate, but a proscription against public use precludes any number of potential conflicts with the refuge system's conservation mission.

Especially on units designated with endangered species protection as a foremost priority, many refuges restrict access either seasonally or in certain areas to protect ecological attributes. The reason for the restriction at M2W sites, however, is very different. The principal reason refuges such as Nomans Land Island, Rocky Mountain Arsenal, and Big Oaks limit public use so severely is not to protect wildlife or plants—though this is a happy by-product that FWS and DOD publications often foreground—but because the places are too dangerous or too contaminated to allow visitors to roam freely. An examination of the site histories reveals that such restrictive policies in fact come from different motives. This deeper look is important to provide the public with a clearer understanding of what processes create M2W sites, as well as to highlight which institutions were involved in creating the hazards that now exist. These historical contexts also illuminate some of the possibilities and limits for future land uses.

In many cases, the FWS provides clear information that the new refuges are not widely open to public use because of military contamination; however, these explanations come in a broader account that highlights the environmental protections offered by such restrictive management. The Nomans Land Island refuge brochure, for example, first turns to the ecological amenities of the site—"these 628 acres of upland and wetland habitat support many migratory bird species including the peregrine falcon during fall migration"—before moving to the more sobering news that "due to its prior use as a bombing range and the possibility of UXO, the island is closed to the public."[10] Framed in this manner, the reading public may be less likely to question why the DOD has refused to clean up even this one square mile of contaminated land and instead can feel confident that the off-limits island serves as a fine ecological sanctuary.[11] This points to one of the less tangible yet significant hazards of such military-to-wildlife conversions: they may foster a blithe public acceptance of these places as havens for wildlife without examining or holding accountable the actions and institutions that produced such damaged landscapes.

This lack of concern or understanding is not usually characteristic of the FWS personnel assigned to manage M2W sites. Most of the M2W refuge managers I've met express a strong interest in overseeing a thorough DOD cleanup of their refuges to ensure the safety of their staff, the visiting public, and the resident wildlife and ecosystems. The catch, of course, is that the military only occasionally shares this same commitment to a complete cleanup, yet it holds the purse strings.[12] At more than one remediated M2W site, FWS personnel are working to assess the long-term physiological condition and reproductive success of wildlife in order to gauge whether lingering military hazards pose a danger to refuge denizens. But most officials seem to agree that even if individual animals occasionally suffer ill effects from military contamination, at the population level the new refuges generally provide usefully secure habitat.

Whether intentional or inadvertent, the recasting of military practices is often facilitated by emphasizing an M2W site's newfound ecological amenities. FWS information about the Shawangunk Grasslands NWR, which emerged from the closure of the Galeville Military Airport in 1994 (the DOD transferred the land to FWS control in 1999), highlights the refuge as "one of New York's top 10 areas for grassland dependent migratory birds."[13] This view reverberates in popular accounts of M2W sites. One author of a newspaper travel piece on the Shawangunk Grasslands refuge reminisced, "As I followed the wildflower-fringed gravel path, I had to remind myself

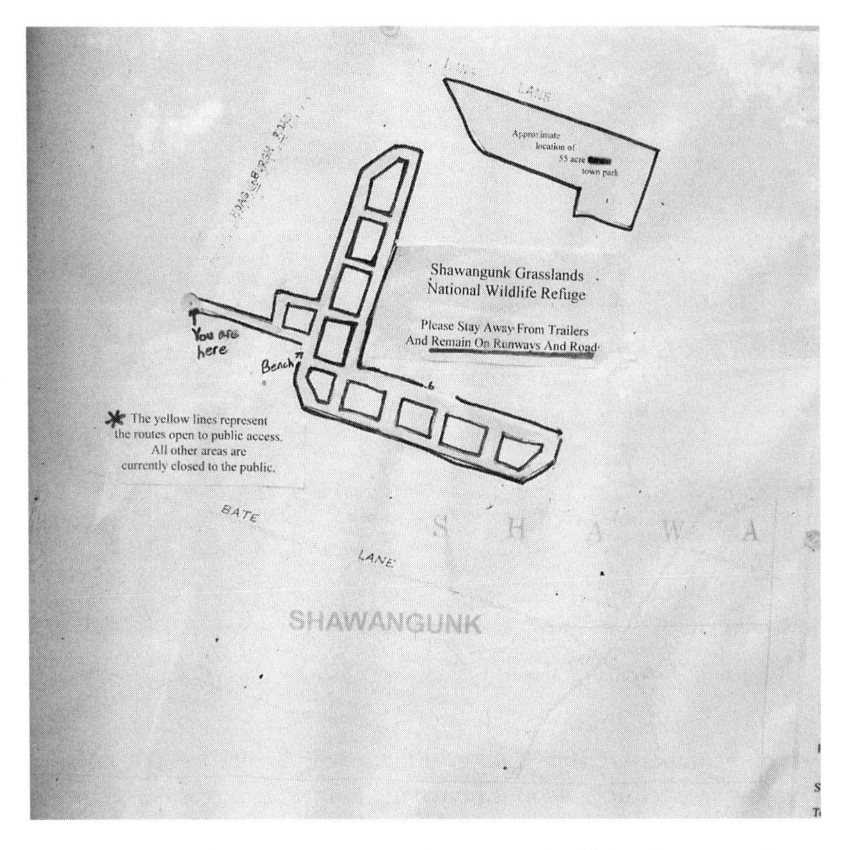

4.2. Diagram of trails, Shawangunk Grasslands National Wildlife Refuge, New York

that the sunny meadow spread before me was once an airport. It was so easy to forget."[14] This description, written just a few years before I visited Shawangunk Grasslands refuge myself, somehow overlooked the fact that the only "trails" at the Shawangunk refuge at the time of the author's visit were mile-long concrete runways, obvious relics from the site's former use as an airfield for army aircraft, special operations drills, and FBI landings (fig. 4.2). When I visited the refuge in 2005, I was struck by the beauty of the site, with mist rising off the meadow grasses, bobolinks flitting through the air, and the Shawangunk escarpment looming in the near distance; I also found it impossible not to notice that I was hiking on a broad expanse of hardened concrete runways (fig. 4.3). (The runways have subsequently been removed at this refuge, so a visit today likely elicits a very different awareness.)

4.3. Runway-turned-trail at Shawangunk Grasslands National Wildlife Refuge, New York

The National Wildlife Refuge System's ecologically focused mission serves as an asset in explaining public use restrictions at the new Shawangunk refuge:

> Unlike national parks, state parks, and state forests, the management priority at national wildlife refuges is "Wildlife First." These lands are managed by the U.S. Fish and Wildlife Service, which is the only agency of the U.S. Government whose primary responsibility is fish, wildlife, and plant conservation. Public uses of national wildlife refuges must be compatible with plant and animal conservation. Our guiding legislation identifies priority public uses on national wildlife refuges that can be allowed if they are compatible with the management of that refuge for wildlife.[15]

Only after these opening explanations does the diligent reader learn that, "Because of potential safety hazards left by the military, *public access is restricted to existing roadway and runways*" (original emphasis).[16] Also coming later is the fact that the ecologically valuable grassland is a by-product of the old military runway: "The grassland that you see today was created when the military filled a wetland with tons of earth to make the airstrip in the 1940's [*sic*]. . . . The grassland persisted over the past 50 years by routine mowing and livestock grazing to remove emerging woody plants."[17]

The FWS can scarcely be accused of deceiving the public here; the agency's publications, exhibits, and employees are quite open about the military past of these new refuges. The overriding story that sweeps across these places, however, serves to emphasize the new management goals of fish, wildlife, and plant conservation. At Shawangunk Grasslands the interested visitor can easily find lists of bird species that frequent the refuge, but will search fruitlessly to find further information on what types of hazards were left behind by the military, or even what military activities took place during more than four decades at the site. In fact, public information is so scant about the military past of this place (it was officially kept as an auxiliary airfield controlled by the US Military Academy at West Point) or the content of its military residues, it's easy not to flinch at the irony of the FWS's parting admonition for refuge visitors to "leave only footprints, take only memories."[18] Only the cynical few may be left to wonder: Was this the guiding principle for operators of the Galeville Military Airport as well?

In terms of the actual management policies or programs in place to encourage or prevent public use, national wildlife refuges created from former military sites don't really stand out. Area or seasonal closures are a feature common to many refuges, though most are for reasons of conservation intent rather than public safety. In order to understand how M2W conversions affect the main agency involved, the FWS, as well as the public, it is important to look more carefully at the reasons these sites carefully limit public access and use. Liability sits high on this list.

As I've already highlighted, the FWS is the most impoverished US land management agency: it receives fewer dollars per acre than the National Park Service, Forest Service, or Bureau of Land Management. The agency's deferred maintenance backlog in 2015 was estimated at $1.3 billion, more than two times its annual appropriation for refuge operation and maintenance.[19] Ultimately, the reason so many M2W refuges currently operate with severe limits on public access or use is not because this is necessarily the preferred way to manage the lands for their conservation mission, but rather because the DOD has not cleaned most of these closed bases to any thorough degree and the FWS has neither the money nor the expertise (in most cases) to conduct the cleanup itself. One FWS official summed up this "definite liability" of receiving military lands for M2W refuges by commenting, "When you arrive as a manager [at an M2W refuge], your first response is, 'Oh my God, why did we take this?'" A refuge official at a different M2W site described the refuge as "amazing biologically," but expressed frustration at how constrained his management options were due to widespread UXO hazards and how little the military had done to clear this ordnance. In his words: "We took it in the shorts" when it came to the army cleanup. At several other sites, FWS employees described feeling marginalized by having positions at M2W refuges, as they often hear comments along the lines of "Oh, you work at one of *those* places . . ."

Learning from M2W Refuges

I first visited Assabet River NWR in Massachusetts on a bright morning in November 2005. I loaded my infant daughter in the car and drove thirty minutes west from Boston to go for a hike. When we arrived at the refuge, a large FWS sign stood in front of a dirt parking lot. Beyond a boarded-up guardhouse and a shiny gate, a kiosk was stocked with brochures and maps. We set off on Trail C, a zigzagging path through the heart of the 2,200-acre refuge.

Our outing thus far could fit any number of similar visits Americans make each year to the National Wildlife Refuge System. For more than fifty years prior to 2000, however, the Assabet River NWR was known as the Fort Devens Sudbury Training Annex and served as a US Army ammunition storage facility. This explains, at least in part, why a walk in the woods at the Assabet River refuge included some surprises. Trail C, for instance, began as a paved two-lane road (perfect for a baby stroller!), complete with yellow median stripe and speed limit signs (fig. 4.4). Where Trail C intersected smaller gravel roads or trails, these secondary routes were posted "No Pedestrians." In late autumn, after the hardwoods had dropped their leaves, I could see a handful of buildings to the side of the road, abandoned, boarded up, and overgrown (these have since been removed). As Trail C wound more deeply into the interior of the refuge, what at first looked like wooded hummocks scattered every few hundred yards revealed themselves as concrete facades of old ammunition bunkers.

Even at that time—just a month after Assabet River NWR first opened for public use—it was a relatively welcoming place compared to many sites of military-to-wildlife conversion. With more than ten miles of trail, the refuge was open to the public every day of the week from dawn to dusk and the requirement to walk only on designated trails was self-enforced and scarcely seemed threatening. Indeed, during that first November hike I saw several carefree families ambling across an open field lightly rimmed with "no pedestrian" signs. Visitors could read that the area was formerly a military site, but there was little hint of danger and no requirement to sign a

4.4. Roads signs along trail at Assabet River National Wildlife Refuge, Massachusetts

liability waiver or check in with FWS authorities (who had no offices on-site and were rarely present). For those who had visited other national wildlife refuges, Assabet River would not seem out of the ordinary. It blended easily enough, ammunition bunkers and all, into its role as part of the United States' largest system of public lands and waters.

I have returned a number of times to the Assabet River refuge since it first opened and witnessed a variety of changes. The guard shack at the entrance is gone, replaced by a paved parking lot, an information kiosk, and trail signs featuring a batch of new wildlife-friendly names: Otter Alley, Kingfisher Trail, and Fisher Loop. A new bike path connects the refuge to surrounding communities, and the site's wide, flat paths (many of them former railroad beds leading to the ammunition bunkers) have become popular hiking and biking routes. As mentioned earlier, the bunkers themselves have become something of a tourist draw, with local historians and refuge volunteers leading regular "bunker tours" that attract visitors by the busload. Since 2010, the refuge has also featured a new energy-efficient visitor center, which includes interpretive displays on the site's history, ecology, and refuge purpose.

In many ways, the changes that have taken place at the Assabet River refuge have made the site more welcoming to the public—they've expanded educational opportunities, with engaging exhibits, public programs, and interpretive signs; and they've enhanced the focus on wildlife conservation while preserving key features (such as the bunkers) of the site's military past. In short, FWS managers seem to be navigating the challenges of preserving cultural and ecological features dexterously. And yet, in surveys I conducted at the refuge during the summers of 2011 and 2012, when I asked visitors to give three words to describe the place, the top three responses were: beautiful (48 percent of all respondents), peaceful or quiet (36 percent), and wild or natural (23 percent).[20] Just three percent mentioned the military or bunkers, while 15 percent gave the somewhat ambiguous reply of "interesting" and "educational." Refuge visitors also provided their main reason for visiting the Assabet River site: 57 percent pointed to recreational activities (hiking or biking), 12 percent came to view wildlife, and 8 percent came specifically to see the bunkers.[21] These results suggest that educational and historical aspects of the refuge attract relatively little notice, while the site's naturalization from military installation to wildlife refuge has been quite thorough.[22]

This brings up the question of what most visitors learn from visiting Assabet River or other national wildlife refuges that have been created from prior activities such as military use. How much does the new designation as a wildlife refuge overwrite prior histories? As I've already pointed out, the Caddo Lake, Rocky Mountain Arsenal, and Big Oaks refuges are not relict landscapes that simply exist; rather, they have been and continue to be actively created and re-created through a variety of actions taking place at these sites, as well as various deliberations about the politics, science, hazards, and meanings that these places evoke. These landscapes are, in fact, the product of a mixed set of processes, which in most cases include the application of eminent domain and the eviction of rural residents, the military activities that ensued, and the decisions to close and convert bases. Many of these actions play critical roles in shaping how people understand and interact with these places as they now exist as valued sites of biodiversity, as contaminated brownfields, and as specific locations where conservation and militarization are seen as compatible or complementary.

One way of trying to understand how M2W refuges serve as new forms of public land is to consider how these sites may create public benefits or risks. How can we best deal with remaining hazards that were created during decades of military or commercial use? Are there risks at these sites that we can now manage in a safe and open fashion? M2W refuges have emerged in recent decades with new names and, increasingly, new reputations, but in fact remain places in transition and with overlapping characteristics of naturalness (often identified by visitors in terms of beauty, peacefulness, or wildness) and risk.

Land Management on M2W Refuges

When the United States designated the world's first national park at Yellowstone, in 1872, it established a set of priorities for management that have influenced federal public land policies ever since. The NPS Organic Act of 1916 committed the United States "to conserve the scenery and the natural and historic objects and wild life . . . as will leave them unimpaired for the enjoyment of future generations."[23] The weighting of this mandate has drawn considerable attention and debate over the years, as it seems to provide a dual mission of conservation and public use. A two-pronged objective stirs little trouble when both ends are mutually compatible, but proves problematic if public enjoyment conflicts with historic, aesthetic, or natural conservation.[24]

In 1997, the National Wildlife Refuge System finally received a mission statement of its own to guide management priorities.[25] The National Wildlife Refuge System Improvement Act sought to rid the refuges of incompatible secondary uses and established an ecological mission for national wildlife refuges: "The mission of the System is to administer a national network of lands and waters for the conservation, management, and where appropriate, restoration of the fish, wildlife, and plant resources and their habitats within the United States for the benefit of present and future generations of Americans."[26]

The final phrase's emphasis on public *benefit,* versus the national parks' public *enjoyment,* gives subtle notice that the FWS should consider public use a management option rather than a fundamental requirement. In this, the agency (and Congress) intended to avoid the dual-mandate problem that has sometimes hampered the NPS. The FWS was even more explicit in its regulations to implement the 1997 act that conservation was not to be subverted to other goals or uses of national wildlife refuges: "The first and foremost goal of the Refuge Improvement Act is to ensure that wildlife conservation is the principal mission of the Refuge System."[27] Despite these efforts to establish more unified management direction with a strong ecological emphasis, the FWS still contends with multiple objectives at many refuges.

Each wildlife refuge in the national system comes with its own establishment document—a presidential order, administrative transfer, purchase, or legislative act—and these in turn include their own particular purposes that must be reconciled with the overarching mission of the National Wildlife Refuge System.[28] Where the respective purposes conflict, the 1997 Refuge System Improvement Act defers to the primacy of the establishment document.[29] This means that the refuge system's ecological mission set forth by Congress in 1997 may be rebuffed or modified where a particular refuge's purposes chart a course other than conservation. At Crab Orchard NWR in Indiana, for instance, the establishing purposes include the development of approximately 1,100 acres for industrial operations—including a General Dynamics defense munitions manufacturing facility—that are widely viewed as complicating the ecological management goals of the refuge system.[30] The seeming incompatibility of wildlife conservation and industrial production is justified by the FWS simply by citing the refuge's original intent: "Industrial operations is one of the legislated purposes of the refuge."[31]

Coexisting Landscapes

The management policies for many wildlife refuges include restrictions on the types of uses permitted, as well as the timing or location of public access. In some instances, even though these restrictive policies may seem to come simply as a result of *natural* circumstances—endangered bald eagles roosting in cottonwoods of Rocky Mountain Arsenal, for example, or the presence of rare grassland communities in northeastern or midwestern sites—the landscapes of these places are also always the product of contested social processes. Grasslands did not just *appear* at the Big Oaks or Shawangunk Grasslands refuges; rather, they were *created* through a series of actions: the shelling, clearing, and burning of agricultural lands following the designation of the Jefferson Proving Ground, for Big Oaks, and the filling, mowing, and grazing of a wetland at Shawangunk. These actions were themselves the result of a series of decisions. The ecological amenities that are highlighted at both Rocky Mountain Arsenal and Big Oaks, now that they are national wildlife refuges, were made possible only through the federal appropriation (condemnation) of these places from their earlier agricultural and residential conditions, and then later through the decades-long manufacturing or testing of military weapons at these sites.

Despite appearances, the deeper histories and politics built into M2W landscapes very much remain responsible for shaping these sites. When visitors today walk through a field tittering with the song of Savannah sparrows at Shawangunk or happen upon otters playing in a creek at Big Oaks, they may easily forget that these places were not always just so. Imagining this often is exactly the appeal of coming to wildlife refuges or other protected lands that can hold our gaze so differently from the urban landscapes we increasingly inhabit, but we deceive ourselves if we slip too comfortably into the "natural" embrace of such places. As geographer Don Mitchell cautions, "Since social struggle is strategic, compromises often gain the appearance of stability: landscapes become naturalized; they become quite unremarkable."[32]

In many respects, M2W refuges actually *do* seem remarkable and may well stand out from other kinds of federal lands by more clearly blending social and natural elements. M2W sites differ considerably from one another, with an array of landforms, land use histories, and habitat types. But each also brings to the fore some element that clears the temptation to view these as wholly "natural" places. At the Rocky Mountain Arsenal, the triple-lined hazardous waste landfills are off-limits to visitors, but still loom as rounded grassy knolls near the center of the refuge; at Big Oaks, UXO and radiation hazard signs caution visitors back to the roadways; the Shawangunk refuge sparkles with the sounds of songbirds, but even today's visitors might notice that new refuge trails are laid atop the broad imprint of those old airport runways. At Great Bay, Assabet River, and other refuges, the lands are dotted with overgrown ammunition bunkers, while the wetlands and forests of Oxbow NWR cannot entirely muffle the rattle of automatic weapons and thumping artillery at the still-active portion of adjacent Fort Devens. From this peculiar mix of features, M2W refuges emerge as a type of hybrid space—seemingly natural places shaped by military technologies and impacts. When considering multiple M2W sites and their characteristics, the greater concern may not be that any particular place will entirely hide its military influences and impacts, but rather that militarization as a broader process becomes naturalized. From these places and their emergent ecological characteristics, we may no longer identify just how broadly destructive—at home and abroad—our preparation and training for warfare, and commitment to national security has in many ways been. Hybridity of landscape, in this context, may well serve to obscure some of the particular components that form the whole.

The Big Oaks NWR Public Use Map

The visitor map of Big Oaks NWR shows a patchwork of access categories across the 50,000-acre expanse of the site, including day use, hunting areas, roads, streams, and closed areas (fig. 4.5). One shaded block that stands out in the north-central part of the refuge is labeled "Air National Guard

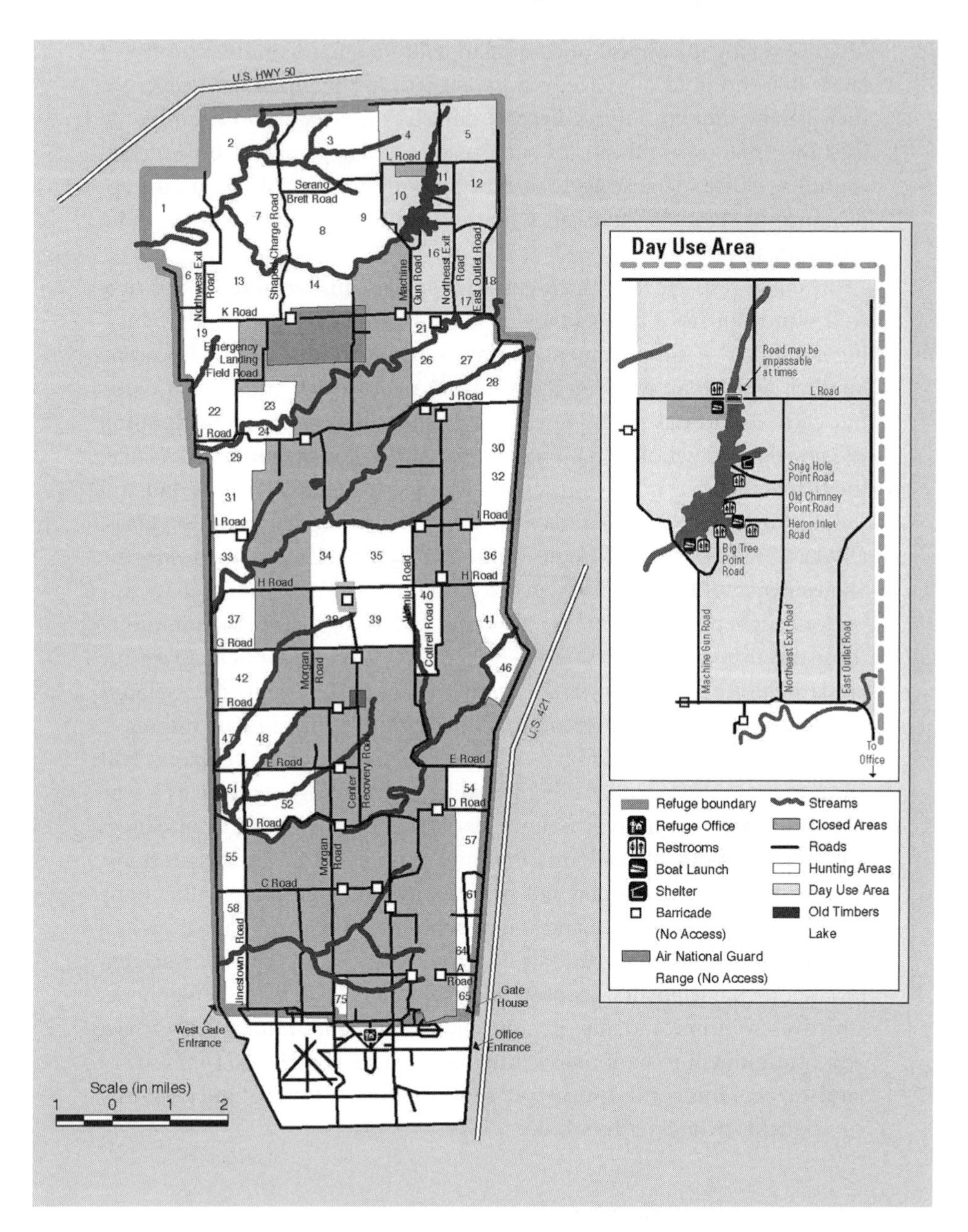

4.5. Public use map of Big Oaks National Wildlife Refuge, Indiana. View the map online at http://www.fws.gov/uploadedFiles/Bigoaksmap.pdf.

Range (No Access)." In the place portrayed by this map, there is evidence both of past productions and current landscapes as they coexist. Indiana Air National Guard bombing runs would seem to be very different in character from FWS activities dedicated to conservation, but these actions take place within the same perimeter fences. Unlike many M2W sites and their residues of past military actions, within Big Oaks military exercises exist concurrently with the FWS's conservation programs (the same is true at the Patuxent Research Refuge in Maryland, which sometimes closes on short notice for national-defense-related drills).

According to the environmental analysis conducted after the army stopped munitions tests at Big Oaks, on the 1,033-acre Indiana Air National Guard parcel, "almost daily, fighter jets come into this area from five states to test shooting accuracy," including air-to-ground bombing and strafing.[33] In its environmental analysis, the FWS labeled this activity "incompatible" with the refuge, and the bombing range was subsequently kept out of the refuge acreage. It remains an inholding near the center of the refuge, however, "until such a time when the range is no longer needed and the land can be transferred to the refuge."[34] In fact, the continued bombing benefits the refuge in certain ways, as the Air National Guard currently maintains all forty-eight miles of perimeter fencing that encircles the refuge, and also covers maintenance and repairs on a portion of the refuge roads. These services represent a contribution of more than $50,000 annually that the FWS likely would struggle to afford on its own.[35]

Thus, although the Air National Guard bombing range is technically not part of the wildlife refuge, the targeting area exists fully within Big Oaks' refuge boundary. Most weeks from Tuesday through Friday, the air guard conducts aerial operations dropping duds at ground targets. This requires FWS personnel to keep out of a one-to-two-mile buffer surrounding the targeting zone for approximately one hour (during hunting seasons this sometimes also requires a rearrangement of assigned hunting units).[36] One or two times per year, the majority of the entire refuge is also closed off to provide a safety fan for a precision-guided range that includes live munitions.

The environmental analysis of the army's closure of the Jefferson Proving Ground and its proposed reuse as a refuge includes only a brief assessment of the consequences of bombing practice occurring inside a wildlife refuge: "Potential conflicts exist between the Public Use Plan and the Air National Guard activities. People using adjoining areas for wildlife oriented activities would be subjected to the noise generated by low flying jets. . . . Future planning efforts will need to look at this issue to determine if relocation of the range to a better location is feasible."[37] The FWS long assumed that the Indiana Air National Guard would terminate its mission at the Jefferson range in the near future, but after nearly two decades, military officials have given no indication that this will be the case. Since the bombing range is not included as part of refuge lands, the FWS holds no real control over the fate of the air guard base and in fact is kept subordinate to the continued military mission. According to the terms of the army's lease to the FWS, "The Air National Guard's range use schedule will have precedence over wildlife management activities."[38]

The Big Oaks refuge is subjected to hazards that go beyond the nuisance of low-flying jets or the regular closure of a portion of its lands. In 1998, an Indiana Air National Guard pilot lost control of his F-16 jet and was forced to eject. His aircraft exploded on impact in the refuge.[39] Such events bring a sobering element of reality to the mandatory visitor training currently in place to orient the public to likely hazards of the Big Oaks site. Air National Guard activities also impact the refuge's nonhuman inhabitants—presenting another potential hazard that wildlife must contend with in addition to UXO and depleted uranium.[40]

Portraying Landscapes

Even as cluttered with information as the Big Oaks public use map may be, it actually represents only a portion of the political and managerial struggle over this landscape that has taken place, and in some cases continues. Maps necessarily both highlight and omit a variety of features, and this visitor map includes elements such as roads, streams, and lakes. The map also marks hunting units and a block designation for day use in the northeast corner that indicates opportunities for public use, while noting barricades and the off-limits Air National Guard range. There is no sign, however, that the areas open for hunting or day use remain cluttered with military hazards, and that hunting exists in these particular units not because they are known to be *safe,* but simply because hunting has taken place here for decades. To the contrary, closures that appear on the map with direct explanation (for example, "Closed—UXO") implicitly suggest that open areas might actually be free from such dangers.

There is also no indication on the public use map of the temporary closures that affect the refuge periodically during aerial bombing practices; no hint that in the late 1990s, FWS Director Jamie Rappaport Clark sought and failed to secure a five-year sunset provision for the Air National Guard base's existence inside the refuge boundaries[41] or that the FWS and air guard ever competed for exclusive control of the entire acreage north of the Jefferson Proving Ground's historic firing line. The map does not offer any explanation *why* public day use is only permitted in one corner of the refuge, or reveal that the FWS sought and failed to secure an army cleanup of the entire northern tier of the refuge. The zone harboring depleted uranium ("Closed—DU") appears static and contained on this map, regardless of the questions that continue to be asked about depleted uranium's movement through biological and physical transport, or the fact that the refuge's Nuclear Regulatory Commission permit remains open and allows for additional radioactive materials to be deposited on-site.[42]

It may be tempting to dismiss such a list of questions and concerns because *of course* a single map can present only a select amount of information about a particular place, and each of these concerns might seem rather mundane. This brings me back directly to Mitchell's point that strategic struggles and compromises can gain an appearance of stability that seems unremarkable. Central to a full consideration of Big Oaks ought to be the message that it remains a remarkable place. If a map serves to naturalize the presence of such extraordinary landscape features as thousands of pounds of depleted uranium or millions of rounds of military explosives, then we may quickly lose track of the meaning of the very processes that produced these characteristics. The public may, in short, grow to understand this place as *simply* a national wildlife refuge rather than also recognizing its lasting existence as militarized space.

Public Hazards and Risk Society

One way to view M2W refuges more fully as places characterized by elements of wildlife conservation as well as militarization is through the lens of "Risk Society" offered by sociologist Ulrich Beck.[43] In describing risk society, Beck points to two stages in modern development. The first is characterized by the maturing of governance at the level of the nation-state and industrialization; the second stage comes as mature states age but then find themselves unable to insure against the hazards created during the process of industrialization. In this second stage, hazards are no longer bounded and controllable, and instead create "a new kind of society and a new kind of personal life" centrally affected by risk.[44]

Beck developed his theory with industrial disasters in mind, such as the 1984 Bhopal chemical plant explosion that poisoned more than half a million Indians with methylisocyanate gas; however, M2W conversions and the opening of militarized landscapes to new kinds of uses also fit this reordering of hazards and the breakdown of seemingly rigid boundaries. Much as military base conversions can open up previously closed places to increased public scrutiny and use, they also release hazards into a more public domain. In some cases—such as contaminated soils becoming airborne or toxins entering groundwater—the hazards have long extended beyond the confines of DOD boundaries, but *information* about the hazards was carefully guarded or simply unknown. In other instances, the opening of military sites to new uses as wildlife refuges actually exposes the public to dangers that had been more limited under military management. One chilling example of this was brought into view when I met with a biologist working at the Upper Mississippi National Fish and Wildlife Refuge's Lost Mound Unit in Illinois, which was created out of a former DOD ammunition testing, storage, and recycling facility named the Savanna Army Depot. In order to conduct a freshwater mussel survey, state researchers were groping through Mississippi River sediment to catch and identify mollusks. One particularly large "mussel," once brought to the surface and washed of mud, turned out to be a hand grenade.[45] Munitions experts later identified the grenade as a "practice dummy," but there is plenty of live ordnance still buried in the river's mud.

In other cases, such as the general access granted to hunters at Big Oaks, the on-site exposure may be very little changed as lands transfer from military control to FWS management, but the public *awareness* of hazards can diminish as the reputations of these places shift from explicitly requiring caution to pointing instead toward conservation and the assurance of safety connoted by the label "refuge."[46] Given the conditions of aging infrastructure at many M2W sites, the historical processes and decisions by which these places were created, and the array of hazards now present, these places in many ways serve as prime examples of risk society. What remains less certain, however, is how this view of M2W refuges as components of risk society can inform public management and understanding. Will these places provide scientists and federal managers with new opportunities to critically examine the relationship between technology, militarism, and the environment, as Beck would advocate, or will the naturalization of these places serve to obscure public understanding in a wash of greening militarization?

No Refuge from Risk Society

According to Beck, a risk society emerges through a process of individuals and institutions systematically taking risks over a period of decades. The ultimate consequences may not be foreseeable or, ultimately, fully remediable.

Military installations such as the former Jefferson Proving Ground (now Big Oaks refuge) present a classic case of this type of activity. The army planned and conducted artillery exercises with little apparent regard for the long-term condition of the place in which they operated, or the prospect that such sites might one day be opened to different priorities incompatible with hazardous military residues. This makes sense within the context of a certain time and institution—during the Second World War, the army's principal concern at the proving ground was to ensure the reliability of munitions it was sending into combat. As military historians of the site are quick to point out, the relatively low rate of misfire by US munitions surely saved many American and Allied lives during the war. An ongoing and unplanned effect of the munitions tests, however, is the ordnance that lies scattered across tens of thousands of acres of former Indiana farmland. The decades of munitions tests served a certain purpose, but left behind a landscape that is now technologically, ecologically, and financially prohibitive to clean up.

According to Beck, as risk societies age, various risks begin to dominate "public, political and private debates."[47] In various ways, elected officials, BRAC commissions, refuge managers, citizen advocacy groups, and local residents now find themselves needing to engage with these hazards that had formerly been hidden or seemed contained. Military sites and their toxic and explosive hazards were produced over a period of decades in restricted areas and are now entering society via shifting land use designations that render the relatively off-limits military installations into a more public domain as wildlife refuges.

Remediating the hazards of latter-day risk societies is an extremely costly proposition. As Beck notes, there is also a certain asymmetry to risks and hazards.[48] Though clearly related, risks and hazards often involve different constellations of people. Military activities over the years involved risks taken by a number of military planners and personnel; today's hazards, however, primarily affect refuge staff trying to manage and clean up their newly acquired sites. As I noted earlier, this shift in exposure actually shows in Environmental Protection Agency standards for cleaning up contaminated M2W sites. The typical EPA standard at these locations is gauged to protect the health of a wildlife refuge worker—a threshold less rigorous than that required for most commercial or residential uses where exposure is expected to be essentially continuous. Even though people coming to visit these lands face a more temporary exposure, and the risks to public visitors is extremely low, in a diffused way they too are subject to hazards created during years of military activity. (Of course, there is also resident wildlife at these sites that receives only occasional monitoring and is not typically covered by EPA standards.) There are also off-site hazards that may be inadequately considered—or impossible to determine—as management activities such as prescribed burning or transporting contaminated soils, physical processes such as wind and water, recreational activities, or biological transport via wildlife may mobilize dangerous substances and disperse them beyond the site boundaries. There's no guarantee, after all, that explosives in the mud at Lost Mound won't be carried downstream by Mississippi River floods.

There is in fact a good deal of uncertainty about how best to deal with problems such as contaminated soils, ammunition bunkers, UXO, depleted uranium, and other military residues that linger at most M2W sites. What the effects of management activities will be—and if enough cleanup is even possible to ensure human safety on (or off) the site—remains hard to evaluate and is disputed. According to Beck, it is not possible in any practical sense to insure against the production of some hazards generated by risk societies: we undertake the actions, such as building a nuclear reactor (Beck's example) or impacting 50,000 acres with tons of explosives and depleted uranium (to use Big Oaks as an example), in a myopic state of optimism or ignorance. At best, we postpone effective understanding of the consequences of such actions to many years or generations hence. Even as science and technology advance over time, however, we may still produce hazards that societies cannot effectively manage or mitigate against.

This latter point emerges clearly at many M2W sites where a common response from federal officials is that complete cleanup simply is not possible. At Rocky Mountain Arsenal, the consolidate-and-contain remediation strategy was implemented in part due to a lack of other attractive options. A number of citizens who participated in the decision-making process for the arsenal's conversion advocated for what they saw as a more complete cleanup, in which contaminated soils would be hauled off-site for treatment or stockpiling in a location removed from the Denver metropolitan area. Federal authorities ruled against this option primarily for reasons of cost and scientific uncertainty: there was no way to guarantee the safe hauling of thousands of truckloads of contaminated material on public roadways or railways to a remote destination.[49] As it turns out, and seemingly providing further support for Beck's thesis, even the alternative implemented at Rocky Mountain Arsenal comes with no guarantee of a permanent, safe resolution. Landfilling toxic materials is at best an imperfect science, and it comes with a long record of failures. A reminder of this rests close at hand, at the refuge's Basin F that leached contaminants into groundwater with disastrous results; when it was completed in 1956, Basin F was lauded for being "leakproof."[50]

Lessons of Risk Society

Despite Beck's concerns that we can never operate with full knowledge of risks and hazards, and that industrial societies tend systematically to produce lasting risks, his theory also makes room for positive change. Rather than encouraging individuals to feel powerless that risk is inescapable, Beck urges a thorough reexamination of how modern societies function: "Where everything is at stake, everything can and must be rethought and reexamined."[51] Instead of complacency in the face of risk society, Beck instead sees this condition as a call to action.

Turning again to the context of military-to-wildlife conservation, changes in land use and designation can open militarized spaces to more transparent and public decision-making processes.[52] This suggests that M2W conversions could contribute to changes in society that extend beyond the mere renaming of military bases to an examination of what kinds of actions occur on military bases, how far-reaching the impacts of these activities can be, and what underlying values are represented or fostered in these spaces. This might seem a tall order, but increased public awareness about what actually occurs on military bases may well come as an essential early step in spurring broader public debate about militarization generally. Unlike the human casualties and environmental consequences of foreign wars, the impacts that we uncover from many domestic military bases strike in many ways much closer to home. This can be rather obvious geographically, as local residents learn about toxic groundwater plumes, airborne pollutants, or other hazards that affect them directly. More broadly, as US President (and General) Dwight D. Eisenhower once noted, "Every gun that is made, every warship fired, signifies in the final sense a theft from those who hunger and are not fed, those who are cold and not clothed."[53] These impacts also come in the form of physically or emotionally injured soldiers returning from combat to communities and institutions that then struggle to support them.[54]

It is also useful to recognize that the risks taken during years of activity at military bases ultimately produce hazards that differ from those that come of war.[55] The impacts generated at many domestic military installations come from activities that fit more easily into the context of everyday industrial production: refueling aircraft, for example, or manufacturing chemicals. Once military bases are repurposed and opened (at least somewhat) to public inspection, we often find that violence has still been done to the land, water, plants, animals, atmosphere, and more, but not directly to humans in the same way it is during warfare. Even as the military remainders found at M2W sites become naturalized and viewed increasingly as part of the landscape, the liabilities that come with these products should not in turn be accepted as *natural* hazards. At some level this surely is apparent: military munitions and groundwater tainted by military chemicals are not normal characteristics of a wildlife refuge.

If M2W conversions are to contribute meaningfully to real changes in land use and a more lasting relationship with the environment, then it will be important to come to terms with the chronic uncertainty of the risks that have been taken in these places rather than simply accept experts' assurances of safety.[56] Signs of this increased engagement with militarized landscapes take various form among the M2W conversion sites, from the work of citizen groups at refuges such as Big Oaks and Rocky Mountain Arsenal, to the enhanced role of public regulatory agencies such as state health departments and the federal EPA, to more visible claims of locals to land at Vieques and Pacific Remote Islands sites.[57] In these ways, even as a view of risk society can highlight how industrialized societies have thoroughly broken their relationship with the environment, some of the responses can help chart a course forward as we seek to identify, repair, and restore damaged landscapes, and by these actions reconnect with these places.

Demilitarized Geographies?

While the M2W conversion process generally opens up previously closed spaces to environmental assessments, public hearings, restoration advisory boards, regulatory oversight, and other forms of public engagement that are often rare on active bases, the increased transparency may be only a passing phase if the resulting wildlife refuge (or other conversion category) remains completely closed to public use. At M2W refuges where there is no DOD-sponsored cleanup, no visitor center, or no public access, actual public engagement may remain at or below levels maintained during military management of such places. The opacity, in other words, may persist even in the absence of an active military.

Such entirely "closed" M2W settings as Nomans Land Island, Virginia's Nansemond NWR, and the Pacific Remote Islands refuges still come with some political appeal—elected officials can point to a new national wildlife refuge established rather than to a contaminated brownfield that was once a military base. Despite such claims, the deeper political opening that scholars often highlight for military conversions may not much apply to these settings.[58] In these cases scientific analyses alone may be incapable of establishing sufficient understanding of M2W sites. The ecological amenities of such places may be exceptionally well protected behind locked gates and fences or in distant island settings, but relying solely on biological markers may reveal very little about the broader contexts and meanings that created these conditions. Conservation biology's opening premise that "biodiversity is good," for example, can lead to an oversimplified but seemingly logical conclusion that M2W conversions are also good if they serve to protect biological diversity.

The complex characteristics of M2W sites call for more multifaceted analyses. While this point may by now seem obvious, I return to it here because of the authoritative role that certain sciences come to play at many M2W refuges, and how this affects public perceptions of these places. Staffed predominantly by experts with training in wildlife biology, conservation biology, or other natural sciences, M2W refuges are typically privileged as sites for biodiversity, prompting day-to-day management to proceed accordingly.

While I support this conservation goal in its broader outlines—such environmental protections surely remain an essential component to maintaining a recognizable and functioning biosphere—I question the assumptions built into an easy embrace of such militarized places as *merely* refuges. It can be tempting to visit a place such as Big Oaks, marvel at the flourishing ecology of the area, and think it would great if there were more places like this, bombed to pieces (and thereby protected) by the army. But surely this is far from an ideal conservation strategy. If we look more fully at the historical contexts of Big Oaks and similar M2W locations, and come to understand the physical and cultural damage brought to bear on these sites, I suspect few of us would accept the costs. We can of course find reason to appreciate some of the qualities that now exist at these sites, but should not lose sight of the other characteristics that remain here as well. M2W refuges by another name are simply "demilitarized zones," which, like the DMZ dividing the Korean Peninsula, remain in a sense protected, but also highly militarized and the product of political conflict, social upheaval, and physical destruction.

Of course, even in active militarized zones such as those that divide the Korean Peninsula, it is possible to find new—and in some cases, bizarre—convergences of militarization and conservation. In sites such as this we can encounter questions of how to retain cultural meaning while also pursuing conservation goals amid political and military change well beyond the borders of the United States. In fact, these issues are perhaps even more pronounced in places where militarization has dominated vast landscapes. Nowhere is this more evident than the former Iron Curtain borderlands of Central Europe, which are now being recast as the Green Belt of Europe. In the next chapter, I explore these and other international sites of change more carefully to consider how meaning, memory, politics, ecology, and militarization can be negotiated in very different and surprising ways.

FIVE

Not Nature Alone

In June 2015, news outlets around the world ran versions of an intriguing headline: "Germany to Turn 62 Military Bases into Nature Sanctuaries for Birds, Beetles and Bats." Photos of the new reserves included images of woodpeckers and eagles positioned against derelict watchtowers, scraps of concrete wall, and strands of barbed wire that still linger along the former Iron Curtain borderlands where the new protected areas were to be designated.[1]

The news stories reflected the irony that these new wildlife preserves could be created from militarized landscapes, even though for several decades military-to-wildlife transitions have emerged as an important trend in conservation and ecological restoration on hundreds of millions of acres from Central Europe to East Asia, throughout the United States, and on islands and marine reserves across the Pacific Ocean. As the photos accompanying the news stories suggested, the conversion of former military sites to new purposes of wildlife conservation can seem both inspiring and perverse. When visitors encounter these lands firsthand, the response is often a mix of surprise and confusion: *this* is what used to be a lethal, highly militarized international boundary?

Across much of its 6,800-kilometer length, extending from the Barents Sea in the north to the Black Sea in the south, these Iron Curtain borderlands were formerly characterized by electrified fences, guard towers, concertina and barbed wire, minefields, concrete walls, armed border patrols, guard dogs, and lethal trip wires. With human communities cleared from the borderlands along a swath extending in places up to ten kilometers, and casualties a regular occurrence, the Iron Curtain earned a fearsome reputation as a trans-European death strip. Throughout the Cold War, the Iron Curtain served as the iconic feature of a divided Central Europe.

Today, these same borderlands include hundreds of protected areas and a series of national parks and biosphere reserves that are collectively known as "the Green Belt of Europe." In 2005, the European Union (EU) formally designated an Iron Curtain Trail as one of Europe's longest bicycle routes (fig. 5.1). It is not simply a recreational path. With the Iron Curtain Trail, the EU aspires to provide a means of "experiencing history," a model for sustainable tourism, and a route that fosters a broader sense of European identity.[2] Communities, nongovernmental organizations, and EU member states increasingly value the lands of the former death strip as open space amenities, a living memorial to the Cold War decades of a divided Europe, and important areas of ecological revitalization, cultural meaning, and sustainable development.[3]

On a drizzly September day, I set out on my bicycle from the Bratislava central bus station, heading northwest up the Danube River along the Slovakia-Austria border that for four decades was part of the world's most extensively militarized borderland. Ten kilometers outside Bratislava, I encountered my first Iron Curtain Trail signpost and turned north onto a network of bicycle paths and farm roads along the Morava River floodplains of the western Slovakia border. As I pedaled through riparian forests and wetlands flitting with waterfowl, it was easy to appreciate the contrast from the nearby bustling city, and to think of the simple beauty of the landscape as natural. To be sure, these features of a seemingly natural and *naturalizing* landscape rest very much at the core of what motivated EU representatives, nongovernment organizations, and former Soviet Bloc officials to start formally protecting Iron Curtain borderlands. The appeal of recasting these lands as Europe's Green Belt also emerges, in part, from its irony. The jarring contrast of the Iron Curtain borderlands from the Cold War era to its more recent "green" identity was also, in part, what prompted me to set out biking in the rain. I wondered: How is it that this heavily militarized, fortified death strip now offers an extensive series of wildlife sanctuaries and an extended route for recreational excursions?

It didn't take long for the contrasts of this transition to jolt me from the peaceful reveries of biking along the Danube and the Morava Rivers. Time and again I encountered the deep ironies of this region—perhaps none more so than just northwest of Bratislava when I rounded a bend to find a medieval tower and the ramparts of Devín Castle. The remnants of this 1,300-year-old fortification reminded me that parts of these borders were delineated and militarized long before the Cold War and the hardening of the Iron Curtain. The more recent history was also evident, though. A large rectangular monument at the foot of Devín Castle commemorated hundreds who had lost their lives trying to cross from east to west during

5.1. Iron Curtain Trail

5.2. Devín Castle and Iron Curtain memorial

the Cold War (fig. 5.2). The confluence of the Morava and the Danube Rivers seemed peaceful on this particular September day, but clearly this had not always been the case.

Over the course of my trip, I encountered dozens of monuments marking the Iron Curtain and recalling the violence it imposed on people and place. These ranged from simple crosses commemorating an individual killed while patrolling the border or attempting to flee, to elaborate sculpture gardens and open-air museums. The monuments invariably pulled me out of the bucolic fields or woods, mountains and streams I was pedaling through to recall that these landscapes were the product not of nature alone but a variety of human social practices enacted over previous decades and centuries.

As other examples from this book illustrate, the Iron Curtain borderlands are not unique in experiencing transitions from militarization to conservation; but perhaps nowhere else are these changes accompanied so assiduously on a broad scale by an effort to commemorate the prior militarized state, and to learn from it. The recasting of the Cold War's extensive dividing line into today's Green Belt of Europe certainly comes with an awareness of opportunistic conservation, much as M2W transitions in the United States emphasize. However, in Europe the initial focus has been layered much more actively with the project of remembering. As Michael Cramer, who ushered the Iron Curtain Trail effort from start to finish, explained his vision when I met with him in Berlin, "We can't only look to nature—that would be crazy. Culture, politics, nature, and history all need to be considered together."[4]

Cramer appreciates the ecological flourishing that now characterizes much of the Central European borderlands, and he has stoked this appreciation by bicycling the length of the inner-German border and writing detailed cycling guides of the entire Iron Curtain Trail route—and still he is careful to attend to the social layers that persist, even those often hidden from view. In the official brochure describing the Iron Curtain Trail and its purpose, Cramer writes that the Iron Curtain is "no longer a dividing line but a symbol of a shared, pan-European experience in a reunified Europe."[5]

The three bicycling guides written by Cramer offer kilometer-by-kilometer directions to navigate the route of the Iron Curtain Trail, but they also consistently provide historical context, interpretation, and cultural insights along the way. In a typical entry, the guidebook describes a picturesque section along the Saale River cycle trail to Hirschberg, then notes, "Numerous houses were destroyed on the Saale after 1945 due to the vicinity of the border. Walls, barbed wire and the death zone influenced life in the city until the fall of the wall."[6] The guidebooks also include a number of photo sets that depict paired scenes, the first with Iron Curtain fortifications or guard towers in place, the second with contemporary images of identical views minus fortifications. These before-and-after scenes quietly serve as vivid reminders of how easily history can be lost from view. From my own experience bicycling through Hirschberg, I pedaled blithely across the Saale River bridge, and only when I glanced back for one last view did I realize that the unfettered bridge I had just crossed was the same one shown in a 1984 photo in the guidebook, replete with vehicle barrier and warning sign (fig. 5.3).

To their credit, Cramer and many others seem keenly aware of this risk of historical erasure as the borderlands shift both in physical character and in reputation. As I observed earlier with Rocky Mountain Arsenal, it can be difficult to find an appropriate balance between embracing the new condition of an ecologically restored (or naturalizing) landscape and keeping track of the important human stories and lessons that these places can bring to us. Given the lingering power of the United States' frontier ideal—a sense of national identity and nature that was forged by the possibility of seemingly endless westward expansion—it may come as little surprise that Europeans have worked more diligently to retain cultural features and memories in their own naturalizing landscapes. The crosses, sculptures, and open-air museums invariably pull visitors back from experiencing the route as a

5.3. Saale River bridge, Hirschberg, Germany, 2013

singularly *natural* place and prompt us to remember that these landscapes are also the product of human activity—with lives lived and lost—enacted over previous decades and centuries.

One challenge to commemorating the social layers of a landscape, however, is that there are typically many stories to tell. In the absence of a single unifying narrative that pulls the many deep social layers of the Iron Curtain effectively into view, visitors to this region are often left to come up with their own explanations for how and why it exists as it does, and what to learn from it. This can be a strength in many respects, and inspires diversity in the ongoing human encounters with places, but it can also become bewildering. Even something as basic as the borders themselves can prove problematic. I was drawn to Central Europe specifically to research the post–Cold War transitions along the Iron Curtain borderlands, but for much of my trip the borders I bicycled along had histories that could just as easily draw attention for events during the First or Second World Wars, prior centuries of conflict along the borders of European monarchies and empires, and even earlier borders dating to the onset of the Iron *Age* rather than the Iron Curtain.

This was made clear not only from the looming stone bulwarks of Castle Devín and others I passed along the way, but even more acutely by barriers of language. On the Austrian and German sides of the border, I managed to communicate reasonably well (I studied German for six years, including a term abroad in college), but in Slovakia and the Czech Republic I was quickly reduced to pantomime. Despite my understanding of the profound dislocations and violence caused by forty years of Cold War separation between countries east and west, time and again the realization came that this was but one relatively brief phase among many centuries of boundary making. With most physical traces of the Iron Curtain now dismantled, overgrown, or lost from view, it may be dangerously easy to let the importance of this barrier and the impacts it wrought fade away as well.

The Central European borderlands differ significantly not just through time, but also from place to place. This diversity shows itself most dramatically in the contrast between the inner-German border that formerly divided Germany into East and West, and the international borders farther south that are now open but remain in place. For many Germans, the Cold War division of the Iron Curtain represented a sharp disruption in cultural identity and established new forms of separation that may not compare all that well to those of, say, the German-Czech, Austrian-Czech, or Austrian-Slovak borderlands. Though Germans had long-established subnational identities as Saxons, Bavarians, Thuringians, or Hessians, prior to the Cold War they had generally maintained common linguistic and cultural identities.

Even language and national affiliations blur in places along the Iron Curtain, as evidenced by the case of Sudetenland. Prior to 1945, the western reaches of today's Czech Republic were predominantly settled by Moravian Germans. Following the Second World War, Czechoslovakia expelled the German population en masse, reducing what had been a 90 percent majority of German speakers region-wide to less than 5 percent.[7] In just three years, from 1945 to 1948, approximately 2.8 million Moravian Germans were evicted and relocated across the border east to west.

The removal of the German-speaking population also resulted in a sudden evacuation of property claims along these border regions that later facilitated the establishment of today's protected natural areas. Within three years of the Second World War's 1945 armistice, approximately 3 million hectares were rapidly depopulated; most of this land subsequently became—and remains—state property.[8] Though Germany and the Czech Republic now maintain an open border as members of the European Union, residues of the Sudetenland mass expulsions remain. Designated initially in 1963 and expanded in 1991, Šumava National Park and Protected Landscape Area covers 167,000 hectares of this area that, in 1991, contained fewer than 2,500 permanent residents.[9] Other traces of the historically strained Czech-German relationship are a bit more subtle: In days of cycling through the Czech borderlands, I failed to meet a single person able (or perhaps willing) to speak to me in German. Whenever I asked, "Sprechen Sie Deutsch oder Englisch?" the reply always came back, even if haltingly, "English is better."

The depopulation of Czech Sudetenland and other borderlands was exacerbated as the fortification of the Iron Curtain proceeded in the 1950s. In order to enhance border security, the so-called Zone A of the Czech side between the actual border and the border fence was completely depopulated with virtually no human activity permitted for a width ranging from two hundred meters to five kilometers. In the slightly less restrictive Zone B, extending inward from the border fence up to an additional five kilometers, local settlements were cleared but select activities such as timber or mushroom harvesting were allowed only with a special permit.[10] According to a 1991 census, the population in these Czech border zones was just 60 percent of the 1930 census levels (the last census conducted prior to the Second World War). To this day, many of the communities that once thrived in this area remain virtually abandoned (fig. 5.4). The land cover of the Czech borderlands also reflects these population declines, with increases in forested areas and open fields, and decreases in arable land that have only accelerated during the post-1990 period as state-sponsored subsidies evaporated.[11]

The exodus of German-speaking Moravians following the Second World War and its lasting contribution to conservation set-asides post–Cold War serve as just one example of how *unnatural* the Green Belt of Europe actually is. To be sure, "nature" is flourishing in the national parks and reserves that now concentrate along the former Iron Curtain borderlands. But much like M2W refuges in the United States, the conditions that allow for an ecological resurgence here have a great deal to do with politics, economics, and the array of land use changes that societal changes have spurred. This reaffirms the importance of Michael Cramer's assertion that we mustn't consider the borderlands in terms of nature alone—we also need to bring the ecological changes into conversation with the culture, politics, and history of the region.

The mass expulsion of German Moravians from Sudetenland highlights as well what is at stake if we lose sight of the historical and social layers here. The Green Belt of Europe is surely worth celebrating in many respects—I trust that few would prefer a militarized death strip over a series of reserves dedicated to recreation and conservation—but even the staunchest boosters of the borderlands' ecological flourishing ought to keep in view the fact that these open lands derive from the displacement of millions of people, the rending of Germany (and Central Europe more broadly), and the loss of thousands of lives along thousands of kilometers of militarized borders.

5.4. Abandoned church along the Czech Republic borderlands east of Dolní Dvořiště

It can be tempting to view military-to-wildlife transitions more simply as a process of natural succession, with an uplifting "nature endures" storyline that suggests the world around us is resilient to the point of rebounding from any insult we can apply. If the most toxic square mile on the planet can be recovered within a span of three decades to serve as viable shortgrass prairie at Rocky Mountain Arsenal, or a vast line across Europe can be celebrated for its ecological promise, then why worry? This is, of course, one of the underlying concerns that motivate historical commemoration along these borderlands and other militarized landscapes: if we don't keep in mind the sacrifices and costs by which these "natural" areas have emerged, we may well find some version of these violent, disruptive histories returning. Put more simply, we may find ourselves hastily building new walls rather than remembering why we sought to tear down those we once had.

Korea's DMZ

The case of the demilitarized zone (DMZ) of the Korean Peninsula offers another useful perspective on fortified borderlands and conservation. Much like the Iron Curtain borderlands, Korea's DMZ has drawn attention in recent decades for becoming a de facto wildlife sanctuary.[12] Widely considered one of the most dangerously militarized borders on the planet today, the DMZ separates North and South Korea with a four-kilometer-wide swath that extends across 250 kilometers of the Korean Peninsula. The zone is kept off-limits through a series of fortifications, watchtowers, armed patrols, and nearly two million land mines.[13] Between the fence lines and fortifications, however, a burgeoning ecology in various form persists, as rare white-naped, red-crowned, and hooded cranes use the DMZ seasonally, and resident species include the Amur goral (a type of antelope), Asiatic black bear, musk deer, Amur leopard cat, raccoon dog, and possibly Siberian tiger and Amur leopard.[14] A buffer zone extending several kilometers on the South Korean side of the DMZ expands the effective area where wildlife habitat has flourished during the past sixty years.[15]

In the late 1990s, a pair of Korean American scientists founded the DMZ Forum in an effort to highlight the zone's potential as a nature and peace park, an effort that has drawn widespread support in the years since, including from eminent Harvard biologist E. O. Wilson and philanthropist/media mogul Ted Turner.[16] Much of this interest comes from the ecological conditions thought to exist within the DMZ (in-depth research within the DMZ has not yet been possible), but this is often augmented by the irony of nature thriving within one of the most dangerous (to humans) militarized regions in the world. Headlines often feature this contrast: "Demilitarized Zone Now a Wildlife Haven"; "Korea's DMZ: An Oasis of Wildlife"; and "Korean DMZ Teems with Wildlife."[17] While the efforts to promote the DMZ's wildlife and habitat features have often been spurred by conservation science, the effort also very much involves politics—not only the necessary and mostly absent international cooperation between North and South Korea, but also from South Korea more independently.

South Korea increasingly presents the DMZ as a tourist attraction, and a number of tourist websites and government efforts rebrand the zone in markedly nonmilitarized terms. In 2012, South Korea sought to rename the DMZ and its southern buffer zone of agricultural lands the "Peace and Life Zone," or PLZ.[18] The Korea Tourism Organization now offers PLZ tours and its website explains, "The name 'Peace and Life Zone' pays reference to the unpolluted natural environment and the people's general hope for the arrival of a new peaceful era to both sides of the border."[19] Although the website acknowledges broad outlines of the DMZ's history, it casts the militarization of the zone very much as *historical*: "The DMZ and its surroundings were once the site of fierce battles during the Korean War, but has recovered from its wounds over the last half-century to become a quiet lush green area inhabited by diverse living creatures."[20] The website of one DMZ tour company encourages prospective visitors to "Explore the Excitement of Silence."[21] The more detailed text on the site directly acknowledges certain aspects of the zone's militarization, but again emphasizes qualities of naturalization and the *peacefulness* of the place:

> As you gaze out upon the DMZ from Checkpoint 3 of Panmunjeom's Joint Security Area, your attention is drawn not to the rare opportunity to peek into mysterious North Korea, the North Korean soldiers perched on the watchtower nearby, or your chances of survival in a sudden (and highly unlikely) re-opening of hostilities. Instead, you're captivated by the supreme tranquility—the quiet, the lush green hillsides, the rare birds swooping into untouched marshlands. Here, at the most militarized border on the planet, you feel completely at peace.

The official website of the (South) Korea Tourism Organization similarly emphasizes a "wildlife first" perspective on the DMZ. Its DMZ Tours informational website opens with the heading "The DMZ: A Historically Rich Border." The site mentions the ceasefire agreement that created a buffer zone between the warring states of North and South Korea, then points out how "restrictions, which have been in place for the last fifty years, have helped the ecological resources in the area to remain in an untouched state. As a result, the DMZ is also a unique natural ecosystem, one that is globally acknowledged for its ecological value." Such representations of the DMZ are not false or entirely misleading, but they work to subvert the militarized features of the zone into a broader casting of "DMZ ecology."[22] The marketing of the DMZ as a tourist attraction has been successful by many measures, with an estimated 180,000 visitors coming annually "to spend a day in the clean air and these open spaces."[23]

The DMZ in this way is presented as a place where, absent human activity, nature is thriving. The site need no longer be seen as land sacrificed to the security ambitions of a divided Korea, or the festering outcome of intrapeninsular hostilities and years of violent conflict, but can rather be valorized (and commodified) as territory affirmatively providing ecological amenities to the region. E. O. Wilson's description of the DMZ as a "Korean Gettysburg and Yosemite rolled together"[24] may be in some ways apt and inspiring, but also seems to disregard the active state of militarization that the zone still endures. The concerted rebranding of the DMZ also effectively serves to redeem it in the interest of business development and tourism. Tourists from around the world now come to the DMZ to pose for pictures in faux North Korean classrooms, complete with framed portraits of Dear Leader Kim Jong Il, scurry through tunnels ostensibly dug by the North in preparation for a broad military assault, buy DMZ-oriented trinkets, and enjoy a theme park named "Peace Land."[25]

No doubt, some of this recasting of the DMZ may properly be seen as aspirational rather than crass propaganda. The Gyeonggi Province government, for example, in 2010 completed a "peace world" hiking path near the DMZ to signify Korea's wish for a peaceful world and peaceful closure to the Korean War.[26] Gyeonggi Province views the trail as serving recreational and educational objectives, with local residents chosen as "guardians" who will oversee sections of the trail and offer lessons on local history, tourism, and national security.[27] Other efforts, such as creating a DMZ ecology and peace park, point to a desire "to transform the DMZ into a place that people from all around the world can visit to wish for peace."[28] Websites such as the Korea DMZ Peace-Life Valley express similar sentiments: "We will try to step towards 'Life Society' by opening the door to peace through the concept of life. This is the only reason for the existence of Korea DMZ Peace-Life Valley."[29] The DMZ Forum and other supporters of a transboundary peace park also point explicitly toward a goal of leveraging conservation interests into political reconciliation, proposing that a North-South agreement to protect the DMZ in lasting form for conservation could serve as a bridge to reconciliation and reunification.[30] Border disputes have been resolved via transboundary peace parks in a number of other locations, including the Poland-Slovakia border after the First World War, and in 1999 along the border between Ecuador and Peru.[31] Noting the potential for a similar effort in Korea, environmental historian Lisa Brady has suggested:

> Preserving the DMZ as an internationally recognized and supported symbol for peace and conservation would indeed be a fitting tribute to the costs Korea and its people suffered as a result of the ideological competition and conflict of the Cold War. It would signify, better perhaps than any other act, that healthy environments can promote healthy human relationships, not only at the individual level but at the national and international level as well. A transboundary peace park in Korea would illustrate the importance of environmental issues in diplomatic negotiations and serve as an example of what can be accomplished if issues other than economic and political gain are considered important parts of the diplomatic equation.[32]

The case of the DMZ highlights how politicized—and important—the greening of militarized space can become, even as these transitions often play out in popular media more simply as ironic examples of nature's resilience. Similar depictions come forward in other parts of the world, where bomb sites, battlefields, or borderlands find new acclaim as natural—or naturalizing—places.

Montebello Islands, Australia

Located in the Indian Ocean some eighty miles off the northwest coast of Australia, the Montebello Islands served as the site of three British atomic bomb tests in the 1950s. Today, the islands are managed as a marine park by Western Australia's Department of Parks and Wildlife. The agency's website unflinchingly pitches the park as an "Explosive Attraction" and outlines camping and fishing regulations before briefly noting a history of atomic weapons testing. In addition to providing cautions for foul weather navigation and remaining alert to cyclone warnings, the website's "Playing Safe" section advises, "As slightly elevated radiation levels still occur at test sites on Hermite and Alpha islands, visitors should limit visits to one hour per day. Do not disturb the soil and do not handle or remove relics associated with the tests."[33]

From the information offered by the state's website, it remains unclear what prospective visitors to Montebello Islands are likely to make of the place. The narrative of the text clearly indicates that nuclear tests occurred here, and that some (undefined) level of contamination remains, but it introduces the islands "with their natural land and seascapes, barrier and fringing coral reefs, wide variety of wildlife and rich maritime heritage" as attractions to divers, anglers, and kayakers in a way that emphasizes the islands as *natural* attractions. The photos that scroll across the website reinforce these associations of a thriving natural place, with swirls of tropical fish, corals, and seemingly pristine beaches and bays. One landscape view hints at a concrete structure atop a small rise—perhaps a bunker or monitoring station mentioned briefly in the text—but it scarcely conjures images of the nuclear fireballs detonated decades ago. During my own visit to northwest Australia, tour operators and local businesses highlighted the pristine conditions found along one of the world's largest coral reef complexes, never mentioning the atomic tests that for decades had drawn attention to this relatively remote stretch of the continent.

The contemporary characterization of this place and its history contrasts markedly from, say, the front page of the *West Australian* newspaper published on October 4, 1952, a day after the first of the Montebello Islands tests: "Britain Tests Her First Atomic Bomb Off W.A. Coast," splashed across the top of the page in all caps, accompanied by a photo of the churning mass of cloud that rapidly shot twelve thousand feet high following the twenty-five-kiloton plutonium-core explosion. The nuclear tests conducted at Montebello Islands (aka Monte Bello Islands) were in fact both far more notable and controversial than visitors to today's marine park might realize. The 1952 blast was Britain's first atomic test and marked the nation's entry into the exclusive nuclear club, which at the time included just the United States and Soviet Union. Two later tests, named Operation Mosaic, in 1956 represented the UK's first success toward developing a hydrogen bomb, sent a radioactive cloud over the mainland of Australia, and spurred a lingering debate over just how powerful one of the blasts may have been.[34] A later 1985 Australian Royal Commission report concluded that "the Monte Bello Islands were not an appropriate place for atomic tests owing to the prevailing weather patterns and the limited opportunities for safe firing."[35] Sources disagree whether the second of two Operation Mosaic tests yielded sixty or ninety-eight kilotons, but in either case it represented the most potent bomb detonated by Britain at the time, and clearly exceeded a fifty-kiloton limit that was in place for British nuclear tests in Australia.

While the significant history of nuclear weapons testing is evident, if only lightly treated, in today's Montebello Islands tourism materials, the role atomic tests had upon Aboriginal lands and people is even less visible.[36] A brief note at the bottom of the marine park website reads, "We recognise and acknowledge Aboriginal people as the Traditional custodians of Montebello Islands Marine Park."[37] At the time of the nuclear tests, Aborigines had not inhabited the Montebello Islands for approximately eight thousand years, but thousands of Australian natives lived relatively nearby—and downwind—on the mainland. The 1985 Royal Commission report found near complete disregard for Aborigines living in these lands most proximate to the test sites, noting "The presence of Aborigines on the mainland near the Monte Bello Islands and their extra vulnerability to the effect of fallout was not recognized by either the AWRE [Atomic Weapons Research Establishment] or the Safety Committee."[38] Elsewhere the report concluded, "Scant attention was paid to the location of Aborigines during the Hurricane test. The Royal Commission found no evidence to indicate that any consideration was taken of their distinctive lifestyles which could lead to their being placed at increased risk from given levels of radiation."[39] Risk factors included "inadequate clothing, ingestion of food contaminated with radioactive material, movement patterns, language barriers (many indigenous people could not read the English warning signs), and general health status."[40]

In the decades since the nuclear tests, families of Australian service members who were stationed near the Montebello test sites have reported high incidences of cancer and early death. A study released in 2006 found increased cancer rates among service members who participated in the tests, and the Australian government subsequently agreed to provide cancer treatments and approximately $70 million in compensation and health care for one thousand surviving veterans.[41] In 1986, Aboriginal populations received just over $300,000 from the Australian government to compensate for radioactivity contaminating their lands, and the British government provided $45 million in the 1990s to remediate the test sites and further compensate Aboriginal populations.[42] Considering the dismal treatment of Aborigines in Australia more generally, it should come as little surprise that their interests carried scant influence in the planning, implementation, and aftermath of nuclear tests. This low status came through pointedly in comments by Sir Ernest Titterton, one of Australia's top nuclear scientists throughout the Cold War and chair of the Australian safety committee monitoring the nation's nuclear tests. Titterton suggested that if Aboriginal people objected to the nuclear tests, they should vote the government out of office.[43] At the time of the tests, however, and as Titterton certainly knew, Aborigines did not hold full voting rights—which, along with inclusion in a formal census, did not come until 1967. Aborigines also represent only about 3 percent of the overall Australian population and scarcely wield a decisive voting bloc.[44]

Bikini Atoll

Of course, the Montebello Islands are not the only remote lands to have endured the impacts of atomic bombs, nor are Australia's indigenous peoples the only to have borne the brunt of nuclear weapons tests. On March 1, 1954, the United States detonated its largest bomb ever, a fifteen-megaton blast (approximately 1,500 times the power unleashed at Hiroshima) that obliterated three islands of Bikini Atoll in the western Pacific Ocean. As geographer Sasha Davis points out in his book, *The Empires' Edge*, the test site at Bikini was made possible only by representing the place as an uninhabited deserted island—a casting made somewhat less tenable by the fact that nearly two hundred Bikinians called the place home. The last nuclear test occurred at Bikini in 1958. During the next ten years, the United States worked to fulfill its promise to repatriate the Bikini islanders. Restoration efforts at Bikini focused primarily on clearing radioactive military debris, stripping the islands of vegetation, and replanting thousands of palm trees and other sources of food to revegetate the islands.[45] Although soil samples on the islands turned up high levels of radioactivity, a US government report concluded, "It was the consensus that attempting to reduce these levels by removing the top layer of soil would destroy the limited agricultural capability of the area, therefore, most such areas were left essentially undisturbed."[46]

By 1968, US officials determined that conditions on Bikini Atoll were safe for islanders to return, and in 1972 a voluntary repatriation of Bikinians began.[47] With contaminated soils still in place, coconut and pandanus palms, breadfruit, and other food sources grown on the islands of Bikini Atoll carried a radioactive burden with them. Within just a few years, urine analyses taken of islanders showed elevated levels of radioactive elements, particularly cesium-137.[48] In 1978, the United States determined that Bikini was in fact not yet safe for human habitation, and all Bikini residents were once again evacuated. Ironically, as Davis aptly notes, Bikini Atoll today is just what military planners imagined it to be in the early 1950s: a deserted tropical island.[49] Analyses reported by the Marshall Islands Dose Assessment and Radioecology Program indicated that for hypothetical populations returning in 2010, full-body burdens of radioactivity would "conservatively" come very close to meeting the Marshall Islands' own standards, and that "resettlement of Bikini Atoll may become much more plausible and cost effective."[50]

In the meantime, Bikini Atoll has been designated a World Heritage Site, in recognition of its international importance as a nuclear test site and its role in the escalation of the Cold War. As the UNESCO website describes Bikini: "the atoll symbolises the dawn of the nuclear age, despite its paradoxical image of peace and of earthly paradise."[51] This view of Bikini as a tropical paradise comes through even more emphatically in enthusiastic travel publications, which often highlight the world-class (and very exclusive) diving opportunities that Bikini now offers. Since 1996, Bikini Atoll has been open for recreational diving tours, in part to fund islanders' efforts to restore livable conditions to the islands. Almost as soon as it opened, Bikini Atoll was anointed one of the world's top diving attractions: by 2000, *Conde Nast Traveler* included it on its list of "top 50 World Escapes" and gushed that few places on the planet "could look more like the Garden of Eden."[52]

Divers heading to Bikini Atoll may have an earthly paradise in mind, but they also tend to choose this destination intentionally for its military relics. Other tropical diving destinations can offer more spectacular marine life and tourism amenities, but nowhere compares to Bikini's "nuclear fleet" of wrecks created by the array of nuclear tests (ships were both used as delivery vehicles for the bombs, as well as to test the effects of the blasts). Tours advertising dives at Bikini market these submerged hulks eagerly ("Bikini Atoll—The Ultimate Wreck Diving Experience!") and often include images of massive mushroom clouds rising from the alluring atoll.[53]

Erasing Damage, Erasing History

Although the imagery of an H-bomb set against a diver's paradise is extraordinary, the conflation of the militarized past and the tourism of the present at Bikini brings to mind other military-to-wildlife or military-to-tourism scenes much like those along the DMZ, the Iron Curtain borderlands, and at wildlife refuges in the United States. At Bikini Atoll, the devastation of the nuclear tests is both part of the appeal (to divers) and the lingering heart of the problem in the failed repatriation of Bikinians due to radiation concerns. At Indiana's Big Oaks the people and the places are very different, but the same broad trajectory is at work: with local residents displaced, military testing leads to widespread contamination, which then both prevents the return of former residents and sets the area apart in ways that allow "nature" to flourish. At the center of both these transitions is an inadequate military cleanup that fails to restore these sites to something even approximating their premilitary condition.

This recalls several concerns about ecological restoration that I raised earlier, including how thorough cleanups can be at severely contaminated sites, how to balance ecological versus public health or other more socially oriented goals, and what ought to count as "authentic" restoration. Whether the changes that occur at a site are initiated intentionally through human efforts, such as remediation projects, or more incidentally as landscapes naturalize, we often at some point face the question of how to commemorate the prior histories of a place. Processes of restoration and naturalization are often viewed as affirming improvements to degraded or heavily impacted sites, and indeed, it's difficult to imagine choosing an Iron Curtain death strip over the more recent Green Belt. But many of these changes in a landscape also come with—or in some cases are predicated upon—a variety of erasures. In the case of environmental contaminants being removed from a site or the elimination of ecological damage through restoration projects, erasure is very much the point. When this happens in places that are significant not just for their ecological characteristics but also for their histories of human activity, then erasure can be a source of real concern.[54]

At places where the impacts or activities of militarization still visibly coexist with new associations of conservation or tourism, we see just a smoothing edge of erasure. For tourists visiting the DMZ or divers exploring wrecks around Bikini Atoll or the waters surrounding the Montebello Islands, there is almost surely a steady recognition that any experiences of nature, solitude, or peace they find are directly linked to the fences, minefields, armed patrols, or bombs that created these unique settings. These jarring contrasts are often part of the appeal that makes such places intriguing.[55] And yet, even in these sites of coexisting militarization and conservation or tourism, the new natures that appear can soften the impact and often dull the power of the historical lessons that might otherwise come through. Historian Astrid Eckert has raised related concerns about the way millions of Cold War tourists to the Iron Curtain served not to disrupt acceptance of this brutal barrier, but instead gradually normalized it as part of the Central European landscape.[56]

Visitors to Bikini Atoll may be struck by the size of the Bravo crater or the devastation of the ruined ships littering the seafloor, but they are also drawn to the place for its crystal-clear waters, undeveloped beaches, and absence of people. Considering this, it comes as no surprise to find diving promotional sites that describe Bikini as "the world's best wreck diving in the world's most pristine tropical marine environment."[57] In a more extended lament of the destruction of global marine environments and coral bleaching, the same author describes Bikini and the broader Marshall Islands as "the last remaining Eden. When I put my head underwater I am relieved to see that everything is as it should be!"[58] Elsewhere, the same website includes a historical account of the nuclear tests and mushroom clouds rising above Bikini. But what many might see as the horror of these tests is translated here as the appeal of this place: nobody else comes here so it remains an unblemished Eden. The violence of America's largest hydrogen bomb test dissolves into a convenient way to create a terrific location to scuba dive.

Visitors to the Korean DMZ can encounter a similar disconnect between the actual violence of the place and its more recent representation as a sanctuary for peace and nature. I spoke with one recent visitor to the DMZ who described a surreal experience of taking a tour bus to the Joint Security Area, then having the tour guide emphasize the area's role as a butterfly and wildlife reserve, with scarcely a nod toward the fortifications and armed patrols lining the border. DMZ tours actually do prepare visitors by providing an overview of the Korean War, detailing security measures still in place (including restrictions on taking photos), and offering views of North and South Korean soldiers, infiltration tunnels, and other features of the heavily securitized border. Considering that tours are also supposed to require visitors to sign a waiver acknowledging the danger of visiting an active warzone, it's difficult to imagine that many visitors could experience the scene at the DMZ and not register some sense of persisting militarization. But the proliferation of peace and nature narratives that now focus on this region, along with the ubiquitous souvenir stands, can serve to diminish visitors' appreciation of the actual violence of this place. As one report cast the DMZ tourist experience: "What they see is more likely to be tacky than terrifying."[59] Many visitors to the DMZ, at best, get something of a mixed message of the area as a dangerous, heavily militarized border that can be experienced safely as a tourist destination where nature, increasingly, is thriving.

The balance of this mixed message is perhaps of even greater concern along the Iron Curtain borderlands in Europe, where the first part—dangerous, heavily militarized border—is no longer in place. What are the risks of losing that portion of this region's history, and how might the legacy of the Cold War's death strip be maintained in a way that can benefit Europeans in the present and future?

With ongoing efforts to expand and solidify the European Union, a number of European leaders remain actively dedicated to advancing the project of trans-European unity and identity. The Iron Curtain Trail sponsor Michael Cramer is an elected member of the EU Parliament from Berlin. He identifies European unification specifically as a goal of the long-distance cycling route. As I noted earlier, his guides to the trail include many details on cultural and historical features along the way. The broader effort to transform the Iron Curtain borderlands into the Green Belt of Europe also attends explicitly to goals of memorialization and memory in addition to conservation. The ongoing European Green Belt initiative describes the region as a "Memorial Landscape" and takes care to position its work as centering not just on ecological goals, but also some that are historical and humanitarian:

> The European Green Belt is an exceptional symbol of European history. This living memorial reminds us of the peaceful overcoming of the Cold War and the Iron Curtain. It is a physical reminder in the landscape of the turbulent and often tragic history of the 20th century. . . . One main aim of the European Green Belt initiative is to preserve this lifeline as a memorial landscape. Remains of border fortifications (watchtowers, patrol paths, ditches or border buildings) provide a vivid picture of the inhumanity of the border regime. Along the Green Belt you will find many projects and activities dedicated to making history visible and experiential: For instance a map of the region's military legacy, and guidelines aimed at ensuring the safe use and maintenance of military objects in combination with nature tourism in Latvia on the Baltic coast. Or there is the "Experience the Green Belt" project along the inner-German Green Belt, which makes razed villages visible again and works with students to collect oral history by interviewing contemporary eyewitnesses.[60]

The commitment to retaining physical remains of the Cold War militarization of Central European borderlands comes in striking contrast to military-to-wildlife transitions in the United States, where most historical objects from prior periods of military use are eradicated as a first stage of ecological restoration and repurposing. Efforts to retain or elevate historical layers along the Iron Curtain may provide instructive examples to these sites, but also can inform how similar processes or eventual transitions could play out in areas of active militarization such as Korea's DMZ, where unification or more formal commitments to conservation may yet occur.

Commemorating the Iron Curtain

The organizers of the European Green Belt initiative, the Iron Curtain Trail, and a variety of regional and local efforts each work through websites, publications, conferences, tours, tourist promotions, media events, and other measures to highlight the cultural and ecological changes—and lessons—now found along the borderlands. For those who lived through the Cold War period of a belligerently divided Europe, or share family histories that made the division of Europe personal, the legacy of the Iron Curtain might seem plenty secure. I keenly recall my own relatively easy passages in and out of Czechoslovakia and East Germany as a student-tourist in the mid-1980s. Certainly those with truly hazardous escapes from east to west across the Iron Curtain will not forget the violence and terror that it presented. But my own more recent experiences along these now-opened European borders highlight how easily meaning can slip from a landscape. It may be relatively harmless if the occasional bicycle tourist lapses into easy reveries and simply experiences the beauty of nature reserves and ecological (and cultural) awakenings while pedaling along the Iron Curtain Trail, but there are clearly more menacing sociopolitical currents that impact this region that lessons from the past could productively inform.

Even as I write, international headlines feature waves of migrants surging across Eastern Europe toward the west. The calls of "never again" in response to a divided Europe already sound painfully distant as Hungary scrambles to build a barrier fence along its southern border to keep desperate migrants at bay.[61] The irony of this comes stronger still if we recall that in 1989 it was Hungary that made the first dramatic breach in the Iron Curtain when it opened its border with Austria.

Given current events in Europe, or similar wall-building efforts along the US-Mexico border and in Israel,[62] the history and meaning of the Iron

Curtain fortifications ought to press upon us even more pointedly. It seems well worth asking: How can Europeans keep the history of the Iron Curtain meaningfully in view? And can this be done in ways that convey the violent disruption of militarized borderlands while also making room for new stories and meaning to unfold? Building from my own experiences along these borderlands, I suggest three broad ways this may be done: through texts and narrative accounts, physical traces, and artwork.

Writing the Iron Curtain

Written accounts of the changing character of the Iron Curtain borderlands come in many forms, ranging from Michael Cramer's detailed bicycling guidebooks to newspaper, website, and magazine articles to a variety of signs posted along the roads and cycling routes that crisscross the region.[63] Most all of these discuss the Cold War barriers of the borderlands, the violence of this separation, and the subsequent renewal that is occurring through recreation, tourism, and conservation initiatives. Perhaps the most obvious textual reminders of the Cold War history that help define the Central European borderlands are the large brown road signs posted at most every road crossing of the former inner-German border (fig. 5.5).

The signs feature a map of Europe with the Iron Curtain as a serpentine line splitting west from east, most prominently through a lighter-shaded Germany, and a brief explanatory text (in German): "Germany and Europe were divided here until 8:00 am on 21 December 1989." Each sign attends to the specific time and date when the inner-German border opened in that particular place, and in many cases the seamless contemporary landscape makes it difficult to imagine a lethal series of fortifications disrupting the view. The contrast between present and past can be jarring, but in this way also carries a reminder of how readily history can be obscured, and how rapidly landscapes and memories can change.

Physical Traces

Physical reminders of the Iron Curtain come in a variety of shapes and sizes and also reveal a variety of approaches to maintenance and neglect. Dozens of guard towers still poke from hilltops and fields to mark the former line of securitized land. Many of these are gradually falling to ruin, with peeling paint, broken windows, and boarded doors; but others have become tourist destinations or found new life as bed-and-breakfast lodgings

5.5. Inner-German border

or restaurants with a view (figs. 5.6 and 5.7). Although scholars debate how these different treatments may provide powerful experiences or more effectively secure a sense of lasting meaning and memory,[64] the towers are surely both familiar and distinctive enough that few travelers can pass by without registering their significance as former sites of a watchful, militarized gaze.

The guard towers are joined by more subtle traces on the land, including concrete or wooden border posts, concrete-slab patrol roads, derelict (or, in some cases, refurbished) buildings and towns, and linear swaths of once-cleared land that now shade different hues of green with early successional shrubs and forests. A number of borderland communities have erected open-air museums to commemorate and interpret the Cold War period and the fortifications that gave the Iron Curtain physical form; some of these sites have become tourist attractions in their own right. Mödlareuth, Germany, for instance, has a carefully maintained Iron Curtain museum, and its poignant history as a divided village attracts busloads of school groups and tourists, eager to witness and learn from the enforced division this small village endured (Mödlareuth earned its nickname, "little Berlin," from the concrete barrier that split the town in two) (figs. 5.8 and 5.9).

5.6. Abandoned guard tower, Iron Curtain borderlands, Germany

5.7. Refurbished guard tower, Iron Curtain borderlands, Germany

5.8. Open-air museum, Mödlareuth, Germany

5.9. Open-air museum, Mödlareuth, Germany

Artwork

The meaning and interpretation of the Iron Curtain certainly varies from visitor to visitor depending on one's background, one's personal connection, the sites or seasons of the visit, and many other factors. Even as some messages seem to ring out loudly of the many possibilities of the Iron Curtain ("never again" perhaps comes through most consistently), it likely makes sense also to accommodate a range of responses to this landscape. Texts and physical remains can do this in ways that effectively convey the violence of militarized borderlands, but artwork may offer the strongest mix of memorializing while also making room for new stories and meaning.

In my own travels along the Iron Curtain, I was most struck by the distinctive uses of art to prompt different forms of meaning, commemoration, and interpretation. As with texts and physical remains of the militarized past, artwork takes many shapes and sizes. Some of it comes with signs, names, or other forms of interpretation as well, but many works are simply cast onto the landscape where they can be encountered openly and understood uniquely. In an effort to offer a sense of how art can—and does—function in this way along the Iron Curtain borderlands, I briefly highlight three encounters with artwork along the Iron Curtain Trail.

Austria–Slovakia–Czech Republic Triple Border

The triple boundary marking the intersection of Austria, Slovakia, and the Czech Republic scarcely qualifies as a tourist hotspot. The confluence of two rivers, the Thaya and the Morava, creates a narrow triangle of Czech Republic wetlands, wedged by the forests and fields of Slovakia and Austria. To reach the triple border, I biked a modest paved path along an inconspicuous dike, then turned down a dirt two-track that brought me to the river confluence. Scattered fishing shacks graced the banks of the rivers, and a metal sign identified the location of the three national boundaries. A few meters away, affixed to a weathered concrete block, stood two iron legs, cast from the knees down, bound by a rusting metal chain twining around the ankles (fig. 5.10).

The sculpture came as a surprise and was stirring in its simplicity. The lack of signs or labels (other than a graffiti stamp touting a popular Slovakian television show: *Vo stvorici po Slovensku*) left the sculpture open to interpretation and questioning. What did it mean? Was it a Cold War–era act of defiance, depicting shackles holding Eastern Europeans in place? A post–Cold War remembrance of oppression? A celebration of rusting chains soon to be broken, or was the rusty chain simply a product of time and neglect? The placement and character of the sculpture calls into view what

5.10. Sculpture at Austria–Slovakia–Czech Republic triple border

art historian Miwon Kwon describes as the broader connections that public art can foster: "The site [of public art] is not simply a geographical location or architectural setting but a network of social relations, a community."[65] This, in turn, gestures toward some of the impact and value of public art more broadly in these deeply political, historical, and militarized settings: it fosters questioning, connections, and meaning, but without circumscribing these narrowly. If effective, public art can open these landscapes up to new experiences that conjure past conditions even as new contexts and conditions emerge.

Eußenhausen-Henneberg, Germany

The fluidity of art and its meaning struck me again some seven hundred kilometers farther along the borderlands, at the crest of a hill above Henneberg, Germany. The once-fortified border separating Bavaria (west) and Thuringia (east), Germany, is still marked here by a derelict guard tower, which looms above a sculpture park and eight-meter-tall "Golden Bridge of Reunification." The "bridge," which spans neither road nor river, is the work of Berlin artist Jimmy Fell, who intended to create a monument to German unity rather than division.[66] Yet inside the bridge's sizable arch an array of objects and smaller sculptures bring to mind glimpses of the trauma and terror of the former Iron Curtain: a pair of crutches, the image of a splayed and bleeding body, a painting of a rifle, images of candles, melted sculpted candles, actual wax candles, and paintings of candles with various inscriptions, including one decrying the role of East Germany's former head of the secret police (the Stasi), Erich Mielke.

In the field surrounding the bridge, a number of sculptures call forth select histories of the area, most poignantly the random and abrupt early-morning eviction of residents of nearby Henneberg in 1952's "Operation Vermin." The sculptures here bear small placards with the title and a brief description of each work, but there is still ample room for individual interpretation and for visitors to encounter the site on their own terms. One work, for example, consists of a pile of red-painted stones out of which juts a red flag. The accompanying sign implores, "Don't forget us! 100 million victims of communism," and later, "Add some stones!" (fig. 5.11). Another set of paired signs reads (in German), "You can evict a man from his home, but you can't take the home out of the man." Yet another sculpture seems to bear silent witness to the jarring displacement of Germans from these

5.11. Henneberg sculpture, Germany

borderlands with the stark empty frame of a house placed next to a chair knocked on its side. Visitors are asked to look at these works, braced against an empty sky and a swath of cleared borderland steadily reforesting, and imagine.

Gompertshausen-Alsleben, Germany

On a different day, biking along the Thuringian-Bavarian borderlands, the Iron Curtain Trail steered me out of the small town of Zimmerau and onto a series of forest paths straddling the border. Signs for hiking and cross-country skiing trails dotted the route. It was peaceful here, and I lapsed into daydreams of winter days skiing through the woods. As I emerged into open fields, however, I came to a paved road that ran between the villages of Gompertshausen and Alsleben. At this inner-German border, a large boulder and a bench sit at the base of a towering (5.4 meter high) steel cross (fig. 5.12). Two small plaques are affixed to the cross. The first notes that the sculpture was designed and built by a resident of nearby Alsleben, in the former East Germany. The second is a dedication: "1945–1990. To Remember Those Killed on the Border. To Warn the Living for the Future" (my translation).

5.12. Cross near Alsleben, Germany

The cross is stunning even from a distance: its mesh steel frame appears both transparent and luminescent reflecting the sunlight. But up close, the cross takes on new, more poignant meaning, its beauty tempered with pain. The fencing used for the Iron Curtain was specially designed to thwart climbing and escape, with small, diamond-shaped beveled openings too tight for footholds and edges too sharp for fingers. This stout, cruel mesh was the material chosen by the Alsleben sculptor to fabricate his cross. I think of him, perhaps roaming these very fields, gleaning fence line in the winter of 1990 rather than skiing the forest trails, and I am washed by emotion. The monument is an apt reminder of the promise of rebirth (the border fence, resurrected!), the hazards of forgetting, and the many layers of meaning that emerge from a single long line drawn on a map.

Visitors along today's Iron Curtain borderlands will, of course, each take something different away from their time and their various encounters with this place. For many of us, the conditions we find here bring into view ideas, emotions, and understanding that mix the personal and the political, the contemporary and the historical. The changes that have occurred throughout much of Central Europe are profound in terms of land use changes and ecosystem dynamics, so much so that in just a few decades it seems a real possibility to forget much of what came before. Without direct efforts to keep some of the history and politics and violence of the decades of Cold War division in view, we might well lose certain lessons from these lands that remain desperately important as new crises—whether of immigration, nationalism, or isolationism—continue to come to the fore.

SIX

Army Green

The US Pacific Remote Islands Marine National Monument is the world's largest marine protected area, covering nearly 490,000 square miles of tropical Pacific waters, seamounts, and atolls. Designated for protection initially in 2009 by outgoing President George W. Bush, then expanded fourfold in size by President Barack Obama in 2014, the monument includes seven different national wildlife refuges scattered across a reach of the central Pacific Ocean larger than the area of Germany, France, and the United Kingdom combined.[1]

A White House press release highlighting President Obama's 2014 actions described the new protected area as "one of the most pristine tropical marine environments in the world."[2] The US Fish and Wildlife Service bills the new protected area as "one of the last frontiers and havens for wildlife in the world and . . . home to one of the largest and most pristine collections of coral reef, seabird, and shorebird protected areas on the planet."[3]

The new national monument and wildlife refuges certainly are remote—Johnston Atoll, the nearest refuge to a major city, sits more than 850 miles of open ocean southwest of Honolulu, Hawaii—but to label many of these areas "pristine" requires either a fantastic dose of optimism or a terrible grasp of history. For nearly seven decades, many of the small islands contained in the new protected area were used by the US military for testing, storage, and disposal of chemical, biological, or nuclear weapons; some of these same islands were created from scratch or expanded to make room for military landing strips or staging military operations. Buried amid the glowing recent announcements of conservation successes here in the Pacific is a legacy of military contamination that ranks among the world's most grievous. After years of abuse and military contamination, the United States' sudden concern to conserve these islands and their marine habitats led some critics to label

Bush's 2009 decree nothing more than a stealth move to solidify US military control in the region under the guise of environmental protection.[4]

One of these sites, Johnston Atoll, had a particularly rough year in 1962. That June, a nuclear-armed rocket fired from the atoll failed just a minute into launch and had to be aborted, scattering plutonium-contaminated chunks of the missile and its 1.4-megaton warhead across the island and nearby marine areas.[5] The next month, another nuclear-armed missile misfired and was detonated on the launch pad, causing severe plutonium contamination and a fire that burned across the island. In October 1962, a third nuclear test had to be aborted, this time aloft, again littering Johnston Atoll with radioactive debris.

While the history of military activity in this "remote Pacific" region ought to render suspect its pristine qualities, the various White House proclamations of bounteous habitat and thriving nature across this region can be substantiated by even a casual tourist. Like many other militarized areas, amid the contamination and violence of weapons tests, the plants, animals, and environmental processes of these sites can, in many respects, be considered thriving. Not surprisingly, the irony of this pairing attracts attention: a growing body of literature now comments on "military environmentalism," "khaki conservation," or "ecological militarization."[6] A central premise of each of these terms is that military testing and activities are presented as compatible with environmental conservation. In its most aggressively charitable form, ecological militarization points not just to the compatibility of militarism and conservation, but claims that conservation successes could only be accomplished thanks to the exclusionary policies of military activity. In this view, the tropical atolls of Pacific Remote Islands remain this undeveloped and awash with wildlife in major part due to the nuclear detonations and military exclusions that kept mundane threats of recreational development, commercial exploitation, and even indigenous occupants out of the way. Military control and contamination thus represent sources of salvation rather than destruction for boosters of ecological militarization.[7]

Of course, it is also possible to acknowledge somewhat more critically a dynamic where military activity serves to protect certain habitats, but at the cost of human rights, environmental contamination, a Cold War arms race, or the historical erasure of the processes of militarization. In this way, ecological militarization could be seen as a legitimate approach to conservation, albeit one that carries with it real risks of also legitimizing militarization. This is more than an abstract concern, as a variety of funding and publicity efforts already clearly work to emphasize the environmental stewardship credentials of defense activities.

Environmental Stewardship and the Department of Defense

The headline of an April 2015 article posted to National Geographic's website hailed, "Bombing Range Is National Example of Wildlife Conservation." The article highlighted the impressive concentration of biological diversity that exists on US military lands, which host more threatened and endangered species "than are found throughout the entire national park system, which has nearly three times more land," and chronicled a journey through old-growth stands of longleaf pine on Florida's Eglin Air Force Base. After touting the role Eglin and other military installations in the southeastern United States have played in recovering the red-cockaded woodpecker, the Okaloosa darter, and other imperiled species such as the reticulated flatwoods salamander, eastern indigo snake, and gulf sturgeon, the author concluded, "As I looked up, two tilt-rotor V-22 Osprey [aircraft] emerged above the treetops and arched down river and out of sight. These impressive metal birds symbolized not just national defense but natural defense, because the journey across Eglin has provided new insight into the rare habitats and biodiversity the military was also protecting."[8]

Taking a more systematic approach to documenting the environmental contributions of military lands, the Center for the Environmental Management of Military Lands (CEMML) based at Colorado State University, consists of "a team of environmental professionals experienced in the conservation and sustainable management of natural and cultural resources on DOD lands."[9] CEMML is supported largely by grants from the DOD and contracts with nearly two hundred biologists and resource managers located either on the CSU campus in Fort Collins, Colorado, or at more than forty military installations across the United States. The center identifies explicitly with the ideas of ecological militarization, noting that "CEMML recognizes that military land use and resource conservation are compatible goals that can be accomplished through the integration of sustainable land management practices."[10] Similar messages come through in articles that label such transitions ("From Bombs to Birds"[11]) and signage at the refuges themselves that point to the shift ("From Weapons to Wildlife"[12]).

Closed and transitioning military sites are not the only ones that emphasize the complementary connections between militarization and conservation. The DOD, its service branches, and many of its active installations also promote the view that military practices are closely bound to practices of environmental conservation and sustainability. After all, as anthropologist Lisa Meierotto points out, "Militarization and conservation may have different motives but they both aim to protect the nation and its resources."[13]

Much like the mix of qualities bound into ecological militarization more generally, the features of this institutional military environmentalism include real accomplishments and a genuine dedication to environmental goals, but also raise concerns about core features of militarism that these actions and programs may serve to obscure or legitimize.[14] Given that the core mission of the US military remains focused upon the lethal application of dominant force worldwide,[15] it would seem wise to examine claims of environmental, social, and economic sustainability very carefully.

In its 2016 budget request for defense environmental programs, the DOD asked Congress for more than $3.4 billion dedicated to environmental restoration, environmental quality, and environmental technology activities.[16] While defense supporters are often quick to point out that the DOD's environmental budget easily surpasses that of any conservation organization—the Sierra Club's entire expenses in 2014, for example, were just under $60 million[17]—a more careful look at how and why the DOD allocates environmental expenditures reveals important details.

The DOD's budget category for "environmental restoration," with a 2016 budget request of $1.3 billion, "addresses contamination from hazardous substances, pollutants, or contaminants" at active, transitioning, or closed installations, and also includes cleanup of munitions-related contamination such as unexploded ordnance.[18] Restoration at these sites, which extend across more than thirty-four thousand locations, typically focuses on remediation and cleanup to reduce active hazards and public safety risks. Groundwater and soil contamination are among the most common and persistent problems this DOD program works to address. In many cases this work is done not by the good graces of the military, but as a result of lawsuits that compel the cleanup. As may already be obvious, the money also simply attends to damage caused by the military itself as a part of its direct training and testing activities.

The DOD asked for nearly $1.9 billion for its environmental-quality activities in 2016, which emphasize compliance with environmental laws and regulations (nearly $1.4 billion), protection of natural and cultural resources ($390 million) and pollution prevention ($102 million). This latter focuses on reducing hazardous discharges and the use of hazardous materials in defense operations.

The third budget category in the military's environmental program, "Environmental Technology," requested $200 million in 2016. This includes an array of funding priorities, ranging from new technologies to assist with groundwater cleanup efforts to "robotic laser depainting" systems and cleaner, more efficient maintenance methods for military aircraft. The DOD casts these projects as often benefiting both the environment and the military mission of the institution.[19]

Working through the DOD's multibillion-dollar environmental program budget, then, affirmative resource protection registers less than 12 percent of total funding, and even these activities are deeply influenced by the overarching "mission first" approach that the military pursues. As the DOD itself states: "The Department supports mission readiness and training flexibility by managing its natural and cultural resources to enable continued access to testing and training lands while complying with existing laws."[20] The vast majority of the DOD's environmental programs budget is, in fact, directed toward repairing some of the damage its operations have wreaked over previous decades, working to comply with basic environmental regulations, and mitigating ongoing impacts of military activities so these can continue with as little future disruption as possible. These budget allocations make clear sense in terms of the logic of this institution—it is tasked with operating and maintaining the nation's military capabilities, after all—but also point to the limits of the DOD's environmental commitments. Ecological militarization, at least insofar as military budgets go, clearly privileges the militarization side of the term vastly more than the ecological.

Once again, this isn't to say that the DOD pays no attention to the environment—certainly it does. But its relationship to the environment is typically characterized by instrumentalism. Military leaders increasingly recognize that it serves their mission-first interests and longer-term budget and political aspirations if they can maintain daily operations without running afoul of environmental laws or public opinion.

The DOD earns its reputation as the most destructive and environmentally consumptive institution on the planet, but at times it can also legitimately lay claim to going beyond the bare minimum to protect the environment and promote sustainability. This link between sustainability and mission capability continues to be updated with increasing clarity in plans and reports issued by the major service branches. In 2004, the Army Strategy for the Environment document announced, "We have learned over the past decades that simply complying with environmental regulations will not ensure that we will be able to sustain our mission."[21] The 2007 "Air Force Operational Sustainability Report" similarly highlighted the linkage between mission and environment: "With unprecedented demands on the critical assets, systems and infrastructure that enable our mission, we must embrace and practice steward-ship [*sic*] of the precious resources entrusted to us."[22]

In 2009, Navy Secretary Ray Mabus announced plans to ensure that half of all energy consumed by the navy, both ashore and at sea, would come from alternative energy sources.[23]

In 2010 nearly thirty major army installations had long-term sustainability plans in place, and a number of installations now have goals to achieve net-zero water, energy, or solid waste programs by 2020.[24] Within each of the military services there are also examples of environmental programs and sustainability efforts that meet and push well beyond compliance. While these efforts are most evident at the installation level, military sustainability efforts exist at multiple scales, including service-level initiatives and the broader recasting of DOD strategies. This broader scope is clear, for example, in military leaders' 2016 Earth Day message: "On this occasion, we renew our commitment to protect the environment and preserve the lands entrusted to the Army—lands that enable our combat readiness and our ability to defend this great nation. . . . We continue to lead the Federal Government in initiatives that sustain habitats and territories."[25]

Some of the more impressive installation-level sustainability initiatives in the DOD can be found within a twenty-minute drive of where I live in Colorado Springs. The US Air Force Academy is an 18,500-acre educational and cadet training complex located immediately north of the city. The site includes more than six million square feet of facilities and educates and trains approximately 8,200 cadets each year. In 2008, the Air Force Academy adopted two ambitious sustainability goals: to become a "net zero" electricity installation, and to be a carbon-neutral installation by 2025.[26] Approximately one-third of the net-zero electricity goal was to be met through conservation, shifts in usage patterns, and energy-efficient infrastructure, the remaining portion would require 11 megawatts of on-site and off-site renewable energy production, including a 6-megawatt photovoltaic solar array.

The Air Force Academy has already shown considerable progress in reducing its energy intensity. In order to meet its 2015 objective, the facility needed to achieve a 60 percent reduction in energy intensity compared to 1975 levels. All new construction on-site is targeted for certification that meets the US Green Building Council's Leadership in Energy and Environmental Design (LEED) gold standards that rate buildings for their energy efficiency and sustainable design features. More broadly, sustainability is being integrated into the curriculum for many of the academy's cadets. The first sustainable design course was taught in 2009, and cadets have developed research collaborations with the National Renewable Energy Lab (in Golden, Colorado), established a chapter of Engineers without Borders and passed LEED exams for new construction. Air Force Academy leaders have also proposed a new Academy Center for Renewable Energy that would provide research and teaching opportunities for cadets and faculty alike.[27] From August 2013 to August 2014, the academy reported a reduction in electrical usage of 3.3 million kilowatt hours.[28]

Located immediately south of Colorado Springs, the army's Fort Carson covers 137,000 acres at its main base and another 237,000 acres at the Piñon Canyon Maneuver Site. In 2006, Fort Carson included more than eight million square feet of building space, serving approximately sixteen thousand active duty military personnel and twelve thousand Colorado National Guard, US Army Reserve, and US Navy Reserve. The total base-dependent population exceeds one hundred thousand, including military personnel, civilian employees, and dependents.[29] As a result of military restructuring and the BRAC Commission–directed closures that began in 1988, Fort Carson added new combat brigades in the early 2000s, and from 2009 to 2014 boosted its building square footage by approximately 50 percent.[30] During the past decade, Fort Carson has established itself as a leader in US Army sustainability efforts. In 2006, Fort Carson was the site of the first building in the army awarded LEED Gold certification; by 2015 it had constructed or renovated a total of eighty buildings certified at LEED silver, gold, or platinum levels.[31]

In 2008, Fort Carson installed a 2-megawatt solar array that at the time was the largest in the army and ranked in the top ten largest photovoltaic solar arrays in the United States. The post now has 4.7 megawatts of photovoltaic solar production spread across ten sites.[32] Along with its recent construction boom, the post has dramatically expanded its bicycle/pedestrian trail system, developed storm-water gardens and permeable pavers in parking lots, and has worked to change base transportation patterns through higher density and mixed-use developments.[33]

Fort Carson has also worked to earn environmentally friendly credentials for its training ranges. Faux Iraqi villages feature dirt blocks made on-site with a compressed soil block machine developed in-house, thus eliminating the need to import cinder block or other building materials. A recent push to expand Fort Carson's Piñon Canyon training site drew fierce resistance from southeastern Colorado ranchers, but the army's management of the existing maneuver site has also won awards both from the army and some nongovernmental organizations for its management and attention to environmental stewardship.[34] More expansively, Fort Carson's 2027 sustainability goals target zero waste, 100 percent of energy needs provided by renewable sources, dramatic reduction in water consumption, and achievements in sustainable transportation, procurement, and training land management.[35]

These performance goals are complemented by process-oriented sustainability efforts to establish environmental protection officers in every unit and ensure all personnel complete an annual environmental and sustainability awareness briefing. By 2015, Fort Carson's efforts resulted in a 16 percent improvement in energy efficiency compared to a 2003 baseline; a 55 percent reduction in drinking water use (2002 baseline); nearly half of all solid waste diverted from landfills; and a free shuttle that provides on-post transit for more than ten thousand soldiers monthly.[36]

At the installation level, Fort Carson has established goals for sustainability—and demonstrated a commitment to achieve these—that compare favorably to similar efforts at even the greenest of municipalities, universities, or other institutions of similar size. Here and at sixteen other installations designated as part of its Net Zero Initiative, the army has tried to establish itself as a leader in terms of renewable energy generation and energy efficiency, water conservation and efficiency measures, and solid waste diversion efforts through reduction of disposable materials, reuse of existing goods, and recycling. Although many of these goals are laudable, and installations such as Fort Carson have demonstrated some progress in moving toward these net-zero targets, a 2016 Government Accountability Office (GAO) analysis of the DOD's Net Zero Initiative overall found no integrated institutional strategy or policy, and that "net zero is not a funded program in any of the services, according to DOD and service officials."[37] DOD officials interviewed by the GAO during its research into net-zero defense programs noted that reaching net-zero goals would be prohibitively costly, but that highlighting net-zero goals could "help generate interest in ongoing conservation and sustainability projects."[38] In other words, military officials view the net-zero programs as serviceable public relations efforts, but "fully achieving net zero is unlikely because, in most cases, it would be cost prohibitive."[39]

This view of net-zero and related sustainability programs has been reflected in my own conversations with army sustainability personnel who describe a frustratingly marginalized position on the installations where they work. Filling primarily civilian roles (though many had prior careers within the military), sustainability officials demonstrate clear pride in the solar arrays, recycling programs, LEED-certified buildings, water conservation efforts, and educational outreach they were conducting, but acknowledge that the military command structure generally shows little interest in integrating sustainability as a daily practice for soldiers. The "mission first" credo of the military is especially apparent in the days prior to large mobilizations and deployments, when soldiers routinely dump valuable electronics and household goods as trash, even though many of these items require special treatment for safe disposal or recycling, and could be easily donated on post for reuse. The command structure of the military could choose to build donation and reuse into its transition process for deployment, but this step is regularly triaged in the interest of gearing soldiers up for combat.

This highlights again the broader fact that military sustainability initiatives are almost always subservient to the military's combat and training mission. This makes sense given the core mission of this institution, but it also throws into question just how much sustainability measures count against the broader and more substantial activities of the DOD. If the army's commitment to turning a brighter shade of green persists only in cases where sustainability efforts save money or carry direct tactical advantages, this scarcely represents a shift in institutional goals or values. Sustainability initiatives instead appear simply as another means to shore up longstanding military concerns with the least cost to military personnel and budgets.

To be clear, human safety and finances of course must remain as centrally important concerns. In fact, the well-being and care of the men and women who serve in the military ought to be kept in utmost regard, even when it comes at high financial cost. Sadly, the DOD's commitment to caring for its personnel isn't always commendable. Keeping in mind sustainability's three-pronged emphasis on environmental protection, economic development, and social equity, examining the military's many environmental, economic, and social impacts raises real concerns about how this institution operates. I need only to look again at my own community of Colorado Springs—or my university campus and classrooms, with our military veterans returning for college degrees—to find ready examples of the toll taken by post-traumatic stress and by the many other challenges veterans face when returning from military training and combat to domestic life. I understand the appeal of casting a newly sustainable, ecological military as the model for today's DOD, but this is not the reality that many returning soldier-students encounter or that reports on soldiers or veterans coming home often reflect.[40] This brings up the key question of how to more fully evaluate what ecological militarization really represents.

Ecological Militarization as a Form of Ecological Modernization

The central idea of ecological militarization is that military impacts and activities can be made compatible with goals of environmental conservation. In this framing, the modernization of today's military brings together new technologies, new methods of training and conducting warfare, and new approaches to procurement and production—in short, new ways of doing the business of the DOD—that work to conserve and protect the environment. In its principal outlines, ecological militarization fits comfortably within a broader policy structure known as ecological *modernization*.

Ecological modernization relies on a view that contemporary environmental crises can be overcome by technical and procedural innovation, and that existing institutions can respond sufficiently to deal with these problems.[41] This requires bringing technological, growth-oriented responses to bear on environmental problems and presents environmental protection and economic productivity not as contradictory goals, as media accounts and political posturing often contend, but rather as complementary efforts toward a unified project of environmental and social advancement. Put simply, ecological modernization presents policy makers with win-win scenarios where economic growth continues and environmental degradation is either avoided or repaired. Not surprisingly, for policy makers and elected officials alike, this can offer a route through industrial—or postindustrial—society that sounds both politically and environmentally attractive. This view, in fact, accounts for a number of the sustainability efforts that in European Union countries such as Germany have worked to decouple economic growth from carbon emissions.

With both ecological modernization and ecological militarization, however, skeptics question whether the responses represent a thorough reworking of existing practices and ideologies or just a superficial treatment of symptoms. Is the new military environmentalism, in other words, a legitimate push to soften the bootprint of military activities, or—perhaps like the US Army's Net Zero Initiative—is the main benefit one of recasting perceptions of the military's impact in ways that spark interest in conservation and sustainability projects in a more limited way?

Ecological modernization's embrace of existing institutions and values leads to a principal critique—that it fails to challenge basic capitalist ideologies and may be coopted into little more than a rhetorical greenwashing strategy. As environmental policy scholar Peter Christoff describes ecological modernization: "There is a danger that the term may serve to legitimise the continuing instrumental domination and destruction of the environment, and the promotion of less democratic forms of government, foregrounding modernity's industrial and technocratic discourses over its more recent resistant and critical ecological components."[42] Against this, political scientist John Dryzek counters that ecological modernization may offer "a plausible strategy for transforming industrial society into a radically different and more environmentally defensible (but still capitalist) alternative."[43]

This tension between ecological modernization as a strategy for legitimizing industrial society and one dedicated to reforming it remains actively contested. It also emerges frequently in the context of military transitions, since the conversion of military lands and postmilitary land use designations can be seen as either superficial, least-cost designations to cover up damage caused by military activities, or as beacons of change signaling a streamlined and environmentally aware DOD.

The common casting of ecological modernization as a win-win scenario emerges frequently in military land use conversions as well, where managerial and technological innovation produce military and environmental "goods." In some cases, this presentation of military-environment compatibility can be seen to emerge from the innermost circles of the US government, as evidenced by then Secretary of Defense Dick Cheney's 1990 comments that defense and environmental goals were necessarily bound together.

One way of interpreting this conflict between modernization-as-remedy and modernization-as-rhetoric is to recognize strong and weak versions of this process. Christoff distinguishes between weak and strong ecological modernization in ways that illustrate just how different these related approaches can be. Weak ecological modernization can serve to legitimize technocratic, narrow, and oppressive systems, while strong ecological modernization may open up democratic processes in diverse, ecologically sensitive ways.

In order to better understand how ecological *militarization* functions in the context of social and environmental processes, it is worth comparing some of the changes currently under way in militarization against these two versions of modernization theory. Drawing from diverse sources that address the relationship between military activities and environmental conservation—including scholarly publications, DOD and FWS agency reports, newspaper articles and coverage in popular media, bulletins issued by conservation organizations, and congressional hearings and legislation[44]—I have identified three main elements that characterize ecological militarization.

First, military practices and environmental conservation appear as compatible activities that provide for the national defense *and* protection of biological diversity. In particular, the streamlining and modernizing of military operations in response to contemporary geopolitics creates new opportunities to protect plants, wildlife, and habitat. Second, existing institutions, current and advancing technologies, and centralized managerial responses can adequately accommodate requirements for cleanup, remediation, public safety, and conservation at military sites undergoing conversion. This suggests that no fundamental institutional transformations are required to create or manage conditions of military environmentalism—there's no need to dismantle the DOD, say, and create a Department of Peace and Environment[45]—but instead the DOD and FWS are equipped to handle land use transitions and their respective responsibilities in these changes. And third, economic and strategic military considerations trigger the initial changes in land management objectives at military bases, but the *nature* of militarized places often influences their reclassification to national wildlife refuges. Whether bases are closed as a result of the BRAC process, legislation, or other means, no one seriously claims that such closures occur first and foremost out of a desire to protect rare plant or animal populations. Once economic or military considerations direct a site into closure, however, then ecological factors are often parlayed to generate support for a wildlife refuge designation.

Based largely on examples from the United States and other highly industrialized countries such as Germany, the United Kingdom, and Australia, ecological militarization fits relatively easily with the development strategies of states able to externalize many of the impacts of military activity. Despite cases in the former Soviet Union, where decommissioning of military sites has taken place due to technological failure, lack of centralized state funds, or the obsolescence of key technologies, many less developed countries find themselves without either the incentive or the means to formally decommission militarized sites. The federal "streamlining" of the military infrastructure in the United States (and of US bases internationally) is largely made possible by three main processes: the expansion of US military bases into the territories of less developed regions such as Central and Southeast Asia; the shifting of militarization to less developed regions *within* the United States, such as the arid and relatively sparsely populated lands of the Intermountain West and Southwest; and the outsourcing of military security, food service, energy and transportation supplies, and other essential services to private contractors such as Halliburton, Blackwater (later XE Services, and since 2011, Academi), Wackenhut (now G4S Secure Solutions), and others. At peak operations in the Iraq conflict, for example, the US military supplied approximately 130,000 troops, while private contractors contributed 50,000.[46]

Military streamlining, and the base closures and conversions associated with this effort, are thus not at all synonymous with a process of demilitarization. The fact that US military expenditures rose 48 percent from fiscal year 2001 to 2007[47]—a time span that straddled the 2005 round of BRAC closures across the United States—illustrates a continued commitment to militarization even as domestic installations closed and faced reclassification. This time period also includes the US military campaigns in Afghanistan and Iraq, plus associated base expansions in these and nearby countries. As I write in 2016, the United States continues to conduct military operations, including remotely controlled drone strikes, in Iraq, Afghanistan, Pakistan, Syria, Yemen, Somalia, and elsewhere, despite President Obama's 2013 turn away from the seemingly unbounded War on Terror that had motivated military policy post-2001.

The many disastrous social, political, and ecological consequences of these wars and combat operations have been well documented elsewhere, but what often goes less noticed is that this period of military expansion and activity internationally fit within a domestic policy agenda focusing on a major restructuring of military assets and strategy. This was directed most actively by Dick Cheney, first as defense secretary and later as vice president, and served as a tangible outcome of his early meetings focused on the "new business" of the DOD.[48] As anthropologist David Vine and others have documented, this military restructuring often came with severe costs to the people and lands of diverse settings internationally, most all of which offered fewer restrictions on environmental impacts, little if any oversight, and less public attention compared to bases and military operations that took place in the United States.[49] For example, one leaked US Army memo acknowledged that open-air burning of solid waste in Afghanistan could cause long-term lung, respiratory, and cardiopulmonary problems for tens of thousands of personnel who had been stationed at Bagram Airfield.[50] This finding came despite years of assurances by the military that open-air burns posed no demonstrable health hazard.

Militarization, Technology, and Science

Similar to ecological modernization's reliance on state or corporate institutions,[51] ecological militarization also operates with what is essentially a technocratic, centralized response to military base closures. The process of base closure often does open up military sites to new levels of public scrutiny—both as a result of requirements in the National Environmental Policy Act to invite public comments, as well as with any subsequent formalization of public use and visitor services—and this can certainly shape cleanup and management actions to varying degrees. For example, at Rocky Mountain Arsenal, the base conversion process very much came out of lawsuits, federal and state intervention, media attention, and citizen activism that influenced the remediation and public uses at that site.

In most every M2W case, however, the ultimate decision-making authority still rests with such centralized institutions as the DOD and, to a lesser degree, the FWS. To point again to the example of Rocky Mountain Arsenal, a consortium of army, FWS, and Shell managers known as the Remediation Venture Office (RVO) wielded the greatest power on a day-to-day basis, with the army the lead authority within that triumvirate. The RVO's power was tempered by the terms of court rulings, regulatory agencies, and legislation, but even these were, of course, institutions of centralized authority: the State of Colorado, the US Department of Justice, US Environmental Protection Agency, and the US Congress.

Much like ecological modernization, the treatments prescribed to assist in the reclassification of M2W sites consistently turn to science and technology rather than citizen participation or local processes. One version of this is characterized by the Big Oaks NWR's policy on the management of depleted uranium: when the local citizens' group Save the Valley raised concerns about the possible effects of burning contaminated areas, the agency dismissed the question with a brief reference to a study conducted at the Department of Energy's Argonne National Laboratory, in Illinois.[52] It is difficult not to view such agency assurances of safety with some skepticism, considering the DOD's checkered past and, in places, continued indifference to the impacts of military contamination and worker safety.[53] A different type of response at M2W sites—but one that also defers to a narrow view of technology and control—is present at sites such as Nomans Land Island or former bases in the Pacific where the new refuges are simply kept entirely off-limits. At many of these locations, the only access permitted is to credentialed scientific investigators who are given safety briefings and escorted by federal officials. In the absence of broader public access, debates about the condition or character of these places and how they ought to be managed in the future are effectively restricted to a select group of researchers.

This is not to say that scientific research is a poor way of understanding these places. Clearly science and technology can offer important insights and tools to assess contamination levels at M2W sites, determine ecological condition and function, map hotspots of biodiversity and/or contamination, provide strategies for cleanup, identify UXO, and more. In most instances, the specialized training needed to identify and clean up hazards at former military sites comes from the same disciplines that informed the production of these hazards, whether these are chemistry, soil science, explosives technology, military science, or nuclear engineering. Even accepting all of this, however, it is important to recognize that by restricting the knowledge base and decision-making authority within militarized landscapes to the same perspectives and kinds of expertise responsible for their construction, we likely do not open these spaces considerably to new politics or values.

The degree to which this maintenance of the status quo prevails depends in part upon how science is managed. If, as some science studies scholars point out, science receives "automatic authority in framing what the issues are," then militarized sites will almost surely remain places largely characterized and controlled by centralized institutions of science and technology.[54] Science can also be applied more democratically, however, serving not as ultimate authority but more openly as a "key resource in public issues."[55] In Aldo Leopold's classic *A Sand County Almanac*, the influential conservationist made a similar distinction between how science can be utilized differently for purposes of conquest or to foster the development of a "biotic citizen."[56] According to the environmental ethic outlined by Leopold, science may serve to increase human domination—what he called "science the sharpener of [our] sword"—or it may function more modestly, as a means to pursue our curiosity—in his words, "science the searchlight on [our] universe."[57]

Reliance on a technocratic, authoritarian form of science often contributes to a belief that there are no limits to what human ingenuity can achieve. Inspiring though this may be in some contexts, it can also lead disastrously toward hubris. The promise of technological innovation can promote a view that effectively encourages carelessness over precaution: if we trust that we can develop technologies that will clean up any mess, then we may be less assiduous about preventing these messes in the first place. This is also the fundamental premise of the restoration thesis brought up earlier: that any damage to the environment we cause can simply be fixed and made whole via ecological restoration.

At M2W sites, this faith in society's long-term ability to repair and restore the environment has often been layered with a protracted disregard for any impacts of military activity.[58] Whether this stems from military officials following orders with little attention to local conditions, a behind-closed-gates attitude that military actions don't need to attend to environmental regulations or common safety precautions, a view that environmental conditions have no bearing on the national security mission of the military, or some combination of these and other factors remains open to debate. However, the consequence of this combination of views is quite clear in the chronic contamination problems that haunt many military installations and their surroundings. The GAO estimates that the DOD has contaminated some 15 million acres of land with munitions, which in 2003 it estimated would cost between $8 and $35 billion to clean up.[59] According to a 2010 National

Institutes of Health report, nearly 900 of the 1,300 Superfund sites in the United States—considered the most hazardous toxic sites in the country—are "abandoned military facilities" or facilities that produced materials or products for military use.[60]

At many militarized sites it has also become evident that technologies do not, in fact, currently exist that can sufficiently respond to the chemical, explosive, or nuclear hazards that still reside in these places. At the former United States Naval Training Range on Vieques, for instance, current technologies limit cleanup to munitions lingering on or near the surface, and to limited subsurface treatments of beaches. UXO located deeper than four feet along the strand or underwater remain in place, awaiting better technologies and a more determined push for full remediation. Whether new techniques will be developed or applied in the future remains to be seen. The fact that the isolation and containment of toxic waste at Rocky Mountain Arsenal is based on a thousand-year planning horizon may be taken as a reassuring sign that regulators take long-term contamination risks seriously, but perhaps also suggests that there is no reasonable expectation for a rapid technological cure to actually neutralize the chemical contamination.[61]

Military-to-Wildlife Transitions and Risk

Recognizing that some technologies simply cannot be controlled over the long term highlights one of the key lessons of risk society that emerges from a number of M2W conversion locations: land managers have discovered that they cannot fully insure against the hazard of chemical contamination, radiation, volatilization of depleted uranium, or UXO deemed too ubiquitous to address with any palatable cleanup scenarios.[62] This brings up an uneasy relationship between the tenets of risk society—particularly the idea that aging modern, developed countries will find themselves unable to insure against the hazards created during their own processes of industrialization—and the more optimistic vision of ecological modernization that development and environmental protection (or health) can be mutually accommodated. At a basic level, most M2W refuges rest on the shaky foundations of risk-society hazards even as policy makers seek to explain military-to-wildlife conversions through the beneficial terms of ecological modernization (or more specifically, ecological militarization). The same holds true for many militarized landscapes more broadly, with the added twist that only by virtue of the military hazards can we now enjoy the twin fruits of advanced, highly militarized societies that also protect ecological values.

In fact, risk society and ecological modernization can be seen as related explanations of contemporary environmental politics and public policy, as both rely on particular constructions of science and technology.[63] Whether these two perspectives can somehow be reconciled to contribute toward a productive, enduring, and environmentally safe society remains to be seen. For this task, a "strong" form of ecological modernization may present a real prospect for change that genuinely addresses environmental problems and conservation, thoughtfully integrates modern technologies, and brings decision-making more fully into civil society.[64]

As I outlined earlier, in its strongest form ecological modernization does not simply accommodate existing institutions and technologies, but rather challenges the assumptions of industrialization to reorient along more comprehensive and diverse environmental priorities. These would necessarily extend across traditional environmental, social, and cultural boundaries that limit weak ecological modernization.[65] In its strong version, ecological modernization manages to operate not in spite of or in opposition to the existence of risk society, but rather as a framework fully informed by this critical view of technology and its responsiveness to questions of risk and authority. Some scholars suggest that a more critical public perception of science and technology may actually facilitate the democratization of technical knowledge. If this were to take hold, science and technology could enter more fully into discussion and use by a more general public, rather than being held apart by a privileged cadre of science professionals. In its most positive version, this could contain the worst effects of risk society and help build toward a more open and humane—if also highly technological—future.[66]

Recognizing, then, that it ought to be possible to work from a position informed for the better by risk society *and* ecological modernization, it's worth taking a look at a specific location of militarization and military transition to see how these relationships play out. Focusing on particular militarized landscapes, we ought to ask: Do military-to-wildlife transitions provide examples of integrative and productive change as these optimistic visions of modernization and risk might expect?

Examples of Modernization and Risk

Among the key characteristics of a more open democratic society envisioned by this affirmative perspective on militarization and modernization are social justice and public participation in decision-making. In the buildup surrounding the Second World War, US military planners carved dozens of military installations out of working landscapes and family farms, often on short notice and with the application of eminent domain. The stories of these dislocations still resonate for many locals in the areas affected, including many sites that have been converted more recently from military use to new purposes as wildlife refuges. In places such as Vieques, Puerto Rico; Commerce City, Colorado; and Madison, Indiana, it doesn't take long to hear about families affected by these fast-moving wartime land claims.

In other places, similar changes came after the war in the buildup to the Cold War or preparing for more recent conflicts in Iraq and Afghanistan. Just outside Limestone, Maine, Loring Air Force Base was created in 1947 as the nation's first operational nuclear-armed site and largest Strategic Air Command base.[67] When Loring closed in 1994 as part of the BRAC process, the DOD offered the site to the Maine National Guard, the National Park Service, and the FWS. Each agency politely declined. Later, the FWS changed its position and accepted the facility, with a stipulation that all military structures be removed prior to transfer from the DOD. After two decades, FWS officials reported that nearly seventy buildings and other military structures had been removed from this site, which has been renamed Aroostook National Wildlife Refuge. Despite these efforts, fifty-seven large concrete igloos (or bunkers) remain in the former nuclear weapons storage site at Aroostook. Many of these structures feature twelve-foot-thick concrete walls designed to protect warheads from Soviet attack. Much like the ammunition igloos at Assabet River, these reinforced military leftovers now attract the attention of military history buffs and refuge visitors. As one refuge official told me in 2012, "Everyone wants to come see the bunkers." A new auto tour route through parts of the refuge now accommodates visitors' desires—as the refuge website highlights, "The three mile [auto] route ends at the historic weapons storage area."[68]

More directly relevant to the refuge's wildlife conservation mission, FWS officials have also been evaluating the concrete igloos' suitability to serve as artificial (and highly secure) bat caves. In light of the continued decimation of North American bat populations from white-nosed syndrome, finding secure hibernation sites that can be decontaminated and kept free from the deadly fungal disease has become a major concern for wildlife officials and bat conservation groups. Starting in 2011, biologists at Aroostook began monitoring humidity and temperature conditions inside the concrete igloos. In 2013 thirty infected bats were brought to an igloo that wildlife officials had retrofitted with various roosting materials and shallow pools for drinking water and to maintain proper levels of humidity. Although only 30 percent of the bats survived their forced migration and winter in their concrete caves, many existing natural sites of bat hibernation across the northeastern United States are seeing mortality at 90 percent or greater. Given the prospect of controlling conditions and decontaminating igloos seasonally, these aging military relics may yet prove to be an important tool in trying to slow the decline and eventually recover regional bat populations.[69]

The story about bats using abandoned military bunkers as a hibernation lifeline at one level serves to support ecological militarization's view of military-conservation compatibility. Using unneeded military infrastructure to advance conservation goals would seem to be a classic win-win—and on the surface, it is. But a more thorough look at the transfer of Loring Air Force Base to become Aroostook NWR, and how that refuge exists today, calls into question the winning strategy of receiving a discarded military installation.

When I visited the refuge in 2008, the visitor contact station (which serves as a mix of refuge offices and a visitor center) was kept open only when volunteers from the Friends of Aroostook group were on hand to staff it. This remains largely true in 2016, as the Friends group keeps a nature store in the office open for four hours each day between Tuesday and Thursday; the auto tour route and hiking trails are kept open more regularly during daylight hours, but are all self-guided.[70] Over time, the Aroostook refuge has been funded to support just one permanent staff position—a heavy-equipment operator. The refuge is officially administered by staff based at Moosehorn NWR, located more than 150 miles away. Although the FWS accepted the air force lands with an understanding that the DOD would remove its structures, agency officials in 2012 described "zero dollars" coming from the air force toward cleanup efforts. The DOD did previously remediate five dump sites on the base that contained radiological materials, but it hasn't subsequently tested or monitored the condition of these locations.[71]

At the time of transfer from DOD to FWS, contamination concerns at the Loring/Aroostook site included these five landfills, three Superfund sites (primarily centered on PCB and chemical contaminants), one National Guard shooting range managed as an inholding, and an area known to contain UXO that had been cleared to a depth of one foot. Although most of these may have little impact on the FWS's ability to manage the refuge effectively for wildlife conservation—the agency is actively researching contaminant loads in raccoon and brook trout populations, as well as deformities in local amphibians—many of these legacies of military use will likely remain in place and in need of some attention for many years to come. Even if the DOD can pitch the M2W transfer here as a win-win outcome of streamlining, some FWS officials view the new refuge as something of an unwanted gift. As one wildlife official told me at a similar M2W refuge plagued by contamination, UXO, abandoned military infrastructure, and a paltry long-term cleanup commitment from the DOD, "You never level the playing field when you're dealing with the army."[72] This sentiment was put a bit more bluntly on a poster that hung from the wall in the refuge office: "Transforming Unwanted Baggage into an Ecological Showcase."

Piñon Canyon and Geographies of Sacrifice

Recent events in Colorado have highlighted in other ways how large a gap can exist between how the DOD views its role as conservation champion and how others do. In southeastern Colorado, just a bit more than a two-hour drive from where I live, an ongoing dispute over US Army plans to dramatically expand its Piñon Canyon Maneuver Site (PCMS) has tested the military's claims of becoming a greener, more ecologically responsible institution. In 1983, through a combination of land purchases and condemnations, the army established the 236,000-acre site to provide additional training land for soldiers at Fort Carson. In 2006, the army announced that it wanted to expand PCMS by more than 400,000 acres to make it the army's largest single training site.[73] The area under consideration for the expansion included more than five thousand people, two towns, and several schools.[74]

Not long after official announcements of the army's interest in an expanded PCMS, a local paper, the *La Junta Democrat Tribune*, published a leaked map that it attributed to army sources. The map showed additional plans for expansion that would stretch the PCMS training area clear to the Colorado-Kansas-Oklahoma border and encompass up to 2.3 million acres. Expansion on this scale would consume the base of the region's cattle industry, hundreds of family farms, more than seventeen thousand residents, the Comanche National Grassland, and numerous historic and archaeological sites.[75] Needless to say, southeastern Colorado ranchers were alarmed by the plans, particularly since the army had assured anxious locals during the 1983 creation of PCMS that they would never again look to expand.

The Piñon Canyon controversy also brought to the fore different versions of military environmentalism, with the region's ranchers claiming the mantle of multigenerational land stewardship, even as the DOD highlighted its own green credentials marked by biological studies of the training area and conservation awards citing the sensitive military management of these lands. A group of ranchers and others opposed to military expansion produced a brochure about PCMS that emphasized ranchers' sustainable approach to range management: "Thus the families that remain [near PCMS] are the hardy souls that care for and cherish the land as a family heirloom to be preserved for future generations."[76] For its part, the DOD gave Fort Carson accolades for its management of the PCMS training lands, including an award for Cultural Resources Management, in 1996, and a pollution prevention award in 1998. The National Wildlife Federation similarly praised Fort Carson's management at PCMS in 1988 with a conservation achievement award "for outstanding contributions to the wise use and management of the Nation's natural resources."[77]

The effort to expand military training lands at PCMS highlights, again, the point that M2W conversions come not as a step toward demilitarization so much as they represent a shifting of military priorities and geographies. The turn toward massive, consolidated bases far removed even from rural populations promises to close military practices off from view ever further. Environmental sociologist Valerie Kuletz describes these types of closed, militarized landscapes that extend across the western United States a "geography of sacrifice."[78] The DOD, for its part, has acknowledged that a shift toward larger, consolidated bases in remote areas will allow it to conduct activities that too often (in its view) attract opposition: "In many geographic regions, the [Armed] Services are constrained in their ability to train because of encroachment near maneuver areas and live-fire ranges. Examples include limits on air operations due to noise, ordnance limits at various ranges, reduced availability of ranges for live fire, restrictions on the use of landing beaches, and pressures from local communities to halt training activities such as artillery firing and air strikes."[79]

The closure and conversion of US military installations that has been pursued systematically off and on since the first BRAC Commission convened in 1988 has, at times, been greeted as an opportunity to open these sites to public scrutiny, more open democratic processes, and in some cases new purposes of conservation. While some of this has come to pass—I likely would not have brought my students to visit Rocky Mountain Arsenal during its era of active chemical production—the geographies of this change remain uneven and inconsistent. Moving ahead, it remains important to look at how militarized landscapes with different characteristics have been opened up to public view and interpretation. This can happen either physically, with cessation of military activities, or somewhat more imaginatively at sites that continue to feel the effects of training, testing, or other forms of military control.

SEVEN

Remembering and Restoring Militarized Landscapes

For nearly two decades, I have been intentionally visiting militarized landscapes across North America, and in parts of Europe and East Asia. At times I bring my kids along. Often, I've collaborated with my wife on projects relating to these complex, layered sites where history and nature defy separation. Our friends and colleagues sometimes tease us and wonder why we'd take family trips to old bombing ranges and chemical weapons plants, rather than just enjoy the mountains and wildlands of Colorado and the Rocky Mountain West.

These are fair questions, and I'm tempted to take them somewhat lightly, as they're meant. But in truth, most of us spend more time than we realize in militarized landscapes. The city where I live, Colorado Springs, is bordered on four sides by military installations, including a major complex bored deep into a mountain of granite. One of our most popular tourist attractions, the US Olympic Training Center complex—just a few blocks from my house—occupies the grounds of the former Ent Air Force Base. Despite the array of active and former installations, the oversized presence of military-related jobs, and Colorado Springs' reputation as a military-friendly community, very few residents here would say that they live in a militarized landscape. And yet, if we apply the definition for militarized landscapes that I offered at the outset of this book—places that have been substantially impacted by military or defense activities—Colorado Springs quite easily fits. The city takes a certain pride in its military connections (in 2015 *USA Today* ranked it the second-best large city in the country for military veterans),[1] and a number of its militarized features are readily apparent. As many examples of military-to-wildlife land use changes from around the world illustrate, however, contemporary and historic impacts of militarization elsewhere are often obscured (or disappearing) from view.

Places such as the DMZ, the Iron Curtain Trail, and M2W refuges provide striking examples of the continued blurring of militarization, conservation, and restoration, but even some of America's most prized natural areas also qualify in more limited or historical ways as militarized landscapes. Yellowstone, the world's first national park, for its first few decades was patrolled and managed by US Army troops in order to create some semblance of protection for the site's geothermal and other natural wonders that were being ravaged by commercial and recreational visitors.[2] Yosemite and other early American parks enjoyed similar military protection until a civilian park service and ranger corps—which itself was explicitly patterned after the military—was finally put in place. Both of these national parks, and other sites such as the Grand Canyon, also depended on a militarized removal of native people to meet nineteenth-century notions of natural landscapes.[3]

It is perhaps not so strange, then, to visit militarized landscapes as part of family vacations or at other times in our daily lives; it's just that too often we never notice that we're doing so. Author Viet Thanh Nguyen invokes the ideas of Reverend Martin Luther King Jr., pointing out how even many of the mundane spaces and actions of our daily lives carry traces of militarization: "These are the places where memories of war belong. Most troublesome is the memory of how it [the US war in Vietnam] was a war that took place not only over there but also over here, because a war is not just about the shooting but about the people who make the bullets and deliver the bullets and, perhaps most importantly, pay for the bullets."[4]

In some cases, becoming more alert to the many activities that contribute to war, and the varied histories and geographies where this occurs, can make these events and landscapes more visible. This, in turn, can promote new understandings about militarization that honor the sacrifices demanded of both nature and culture—or more properly, the blending of these domains. This is the beauty, in a sense, of not forgetting, and I'd like to think that we can learn to embrace restoration in a way that focuses not only on erasing ecological damage, but also on protecting and reintegrating cultural meaning.

If one task that lies ahead is to understand where, how, and why militarized landscapes exist as they do, and to try to keep these sites meaningful even as they naturalize and take on new features, names, and uses, then it will be imperative to look for examples of how to proceed. At its base, my concern—and my fascination—with these places that are significant both for their militarized histories and their militarized ecologies is that they challenge us to think more critically and creatively about the world we inhabit. If we are determined to *avoid* the damaging erasure of culturally significant land use histories in these places, can we do so in ways that *allow* us to erase ecological damage in the interest of restoration and conservation?

Reconstruction and Commemoration in Japan

During the spring of 2015, I lived in Japan for six weeks to start a new research project examining formerly militarized landscapes on the island of Honshu. I planned to visit a number of sites that were clearly associated with military preparation, planning, or impacts, and assess how these places have subsequently been redeveloped or commemorated. Longer term, my goal was to evaluate how these militarized and demilitarizing landscapes in Japan compared with similar kinds of landscapes in other parts of the world, including M2W refuges in the United States and the Iron Curtain borderlands of Europe. I wasn't initially aware of particular sites in Japan that had undergone these same kinds of military-to-wildlife transitions, but I was curious to see how processes of naturalization were taking place in other highly industrialized societies, particularly those with experiences in the Second World War and Cold War that differed from those of the United States.

It was my first visit to Japan, and almost immediately I realized that I would need to reorient my research in fundamental ways. Although I had read quite a bit about Japan in advance of my trip, including useful treatments of the country's modern history, once I set out to visit "militarized sites" in Tokyo Prefecture it quickly became clear that my understanding of the particular military geographies I'd planned to study was not fully formed. At the many transitioning military installations I've visited in the United States, I had found it relatively straightforward to identify the boundaries of an installation. Even along the Iron Curtain borderlands, the zones of historic displacement and fortification were specifically delimited. Often these were marked with signs or fences and labeled even on highway maps. This is not to suggest that twentieth-century militarization in the United States or Central Europe has been limited to discrete locations—it has operated, of course, much more broadly through a variety of sociopolitical, economic, and geographic connections—but in terms of physical landscapes, these military imprints are often fairly well defined.

In Japan, as in many other countries that have borne the brunt of modern military power, the physical presence and impacts of militarization are much more broadly inscribed. (The same is true in many locations of active warfare across Europe and other parts of the world.) There are, of course, particular sites of former military production and preparation such as airfields, manufacturing plants, barracks, and training grounds in Japan that resemble venues with similar purposes in the United States, but the scope of military *impacts*—where places or entire regions were transformed by warfare itself—stands out in Japan. Or, rather, it did for a time. What I discovered, once I arrived in Japan, is that the impacts of warfare affected many parts of the country so broadly and so disastrously that they have become hard to identify today. This sounds counterintuitive, but points to a curious irony of warfare: that the more devastating and widespread the effect, the more difficult it may be to commemorate and keep visible what happened.

The community where I lived in Japan, on the west outskirts of Tokyo, illustrates this paradox. When I first arrived, I was struck by the modernity of the buildings and infrastructure. Initially I assumed this must be relatively recent urban sprawl, and that Tokyo simply hadn't managed to build out this far until recent decades. This seemed odd, as there was a 1,300-year-old national temple just a couple blocks from where I lived. If an ancient provincial road and temple complex could find its way to this area in the 700s, then surely Tokyo-area developers could have found their way here before the 1970s. And of course they did—but prior to that, so did American bombs.

What I learned, and surely should have recognized sooner, was that much of the area in which I was living—and roughly two-thirds of the greater Tokyo metropolitan area—had been destroyed by the intensive incendiary bombing staged by American warplanes, most particularly from March 9 to 10, 1945. In less than twenty-four hours, nearly three hundred B-29s dropped more than 6,500 bombs that built into a firestorm to engulf Japan's largest city, killing more than 150,000 people. Similar degrees of destruction rained down upon most major cities in Japan, leaving vast swaths of the country facing decades of reconstruction.

The extent of the damage was horrific in countless ways, but longer term also posed real challenges for commemoration. Even relatively isolated sites of military impact can create difficulties in terms of how they should be recalled and commemorated, but the task grows dramatically when there is no longer a particular site to mark, but rather an entire metropolitan area. Where would a monument be placed? Should it try to memorialize certain neighborhoods or clusters of bombs, or simply the entire series of attacks and the destruction more generally? Facing a widespread geography of militarization, how could meaning best be linked to place?[5]

I realized as I considered these questions that they reflected similar important queries about meaning, memory, and erasure that I had been working through in the context of military-to-wildlife transitions elsewhere. The specific contexts differ between M2W refuges in the United States and Japan's devastated cities, to be sure, and it would be inappropriate to suggest that the impacts are comparable in a whole catalog of ways, but in terms of trying to figure out how to most effectively maintain meaningful histories and keep these in view despite significant land use changes and shifting identities of place over time, these varied military geographies might help to inform each other.

Many of Japan's militarized landscapes have gone through more than seven decades of change since the impacts of the Second World War (or the Asia Pacific War, as many Japanese refer to the period of active conflict from 1931 to 1945). During this time, most all the urban areas destroyed by Allied bombs have been rebuilt, as Japan was forcibly demilitarized and subsequently reoriented from a position of military power to one of economic production. The most intensively impacted militarized landscapes received a variety of restoration treatments, ranging from reconstruction into expansive public parks such as Yoyogi and Showa Kinen; to commercial and transportation hubs, such as the train station, mall, and Ikea home furnishings store complex at Tachikawa; to a mix of civilian purposes such as the Chōfu airfield, park, and stadium complex (figs. 7.1–7.3).

At each of these sites, it can be exceedingly difficult today to imagine what they looked like in 1945 at their nadir of destruction. Given that two generations have largely passed since this time, and that the direct survivors of the war become fewer by the year, the prospect of a collective loss of memory in Japan has become a real concern. (The same is true in Germany, despite its mandatory inclusion of the Holocaust and Nazi era in its school curriculum.)[6] This has been exacerbated in recent years by nationalist political administrations in Japan that have lobbied to rewrite social studies textbooks—to downplay the role of Japanese aggression during the first half of the twentieth century—and to revise the country's pacifist constitution to allow for an expanded military presence (under the US-drafted constitution post-1945, Japan was allowed only to maintain national "Self-Defense Forces").

My own visits to these places—which I sought out specifically because of their militarized pasts—illustrate how easy it can be to normalize them today as recreational, commercial, or civilian spaces. The playgrounds, water parks, gardens, and grassy expanses that now grace Showa Kinen readily invite new connections to this site that overwrite its history as a military airbase used both by the Japanese and, later, occupying US forces. Dedicated to the Emperor Hirohito (posthumously, Emperor Shōwa), the park is managed as a government memorial to his reign, during which he guided Japan through some of its most internationally bellicose decades, including

7.1. Showa Kinen flowerbeds

7.2. Tachikawa Ikea

7.3. Chōfu airfield and park

its invasion of Manchuria, the bombing of Pearl Harbor, and the Second World War. A visit to Showa Kinen today brings little of this into view, but instead seems to represent a manifestation of Showa's translated meaning: the period of enlightened peace or harmony.

The redevelopment of many of Japan's other militarized landscapes similarly recasts what these sites represent today, and how visitors or residents connect meaningfully with these places. Much like M2W refuges in the United States, the overriding experience at many of these militarized landscapes has become increasingly distant from the sites' earlier emphases on military production, training, or impact. With these previous layers lost from view, the cultural meaning of these places also begins to shift. What we once encountered or recognized as locations of sacrifice, shift instead to become locations of leisure or conservation.

It's tempting to view these kinds of transitions as inspiring. After all, who wouldn't want to see a society rededicate itself from purposes of warfare to purposes of peace? Who wouldn't aspire to enjoy a period of enlightened peace or harmony? The problem with this is not necessarily the aspiration, but the *misrepresentation* that it promotes. Hirohito's period of rule was not, in any genuine way, characterized by peace and harmony (even internally, as multiple Japanese prime ministers and political leaders were assassinated during that time). If the erasure of militarization's physical reminders leads to the enduring miscasting of a society's relationship to the world around it, then this surely is a dangerously high price to absorb. A look at what is likely the world's most iconic militarized landscape may bring this point more clearly into view.

Hiroshima

On the morning of August 6, 1945, the American B-29 *Enola Gay* dropped an atomic bomb that would forever link the city of Hiroshima, Japan, to the dawn of the atomic age. The intensity of destruction caused by that blast, and the uniqueness of the bomb itself, established Hiroshima as one of the world's most infamous militarized landscapes. In this place, military activities and impacts instantaneously became so pervasive that they seemed to define the character of this bustling city.[7] To this day, many visitors view Hiroshima primarily by how it has responded to, rebounded from, or retained the memory of its most dramatic and profoundly destructive wartime event, even as the city has continued to rebuild, change, and modernize in the decades since the bombing.

In Hiroshima, the magnitude of the atomic bombing, both in terms of its destructiveness and its transformation of the global arms race, seemingly guaranteed that this event would be memorialized and not forgotten. After the atomic bombings, global recognition of Hiroshima and Nagasaki quickly changed from simply being places on a map to becoming symbols of their legacy: the global nuclear arms race. Significantly, the memorializations of Hiroshima and Nagasaki were also almost immediately mediated by the United States to represent particular kinds of meaning.

The atomic bombings were—and still remain—highly controversial acts that are viewed very differently depending on one's perception of Japanese and United States imperialism. To many Japanese, the atomic bombs have been seen as American wartime atrocities that went both unpunished and unrenounced in apology. Conversely, to many in the United States, the atomic bombs have been seen as a pinnacle achievement of military technology or, at worst, a necessary evil that ended the war and spared countless lives that would have been lost in a protracted land invasion of Japan. Following Japan's 1945 capitulation, the United States assumed a lead role on the Allied Council that oversaw the occupation and reconstruction of Japan from 1945 to 1952, and in 1948 passed the Hiroshima Peace Memorial City Construction Law that led to the creation of the Hiroshima Peace

7.4. Hiroshima Peace Memorial Park with A-Bomb Dome

Memorial Park. This park now covers a portion of the area most impacted by the atomic bomb near the city's center (fig. 7.4).

Immediately following the atomic bombing of Hiroshima, there was also considerable debate about what to do in terms of rebuilding, with views ranging from almost no reconstruction—in effect, maintaining the destroyed landscape as a type of commemorative mass grave—to removing as fully as possible all traces of the bomb's aftermath.[8] By some accounts, the United States moved quickly to develop the Peace Memorial at least in part to promote the linkage between the atomic bombing and postwar peace, which in turn supports the idea that the bomb was an essential catalyst in securing Japan's surrender and sparing even more lives that would have been lost in a protracted Allied invasion of Japan. Lisa Yoneyama, professor of cultural studies, sums up this policy: "The representation of Hiroshima as the A-bombed city that revived as a 'mecca' of world peace helped disseminate the view that the world's peace and order were attained and will be maintained, not by the United Nations' deliberations or international diplomatic negotiations, but by sustaining the United States' techno-military supremacy."[9]

7.5. Hiroshima plaque with distance to hypocenter

7.6. "A-bombed tree" sign

Working beneath the broad role of Hiroshima's notoriety, one additional factor may have contributed to the success of commemorating this site: the precision with which the location of the bomb's detonation above the surface of the earth—the hypocenter—was known and subsequently marked. Throughout Hiroshima today, the distance from the hypocenter and the prior condition of buildings on the site are marked with dozens of plaques and photos installed across an area ringing ground zero for approximately two thousand meters. Within this radius, even the handful of trees that managed to live through the atomic blast are carefully commemorated with signs denoting their status as "A-bombed trees" (figs. 7.5 and 7.6). At Nagasaki, a black stone monolith now marks the spot directly beneath the plutonium bomb's hypocenter.

These examples illustrate how the restoration and commemoration of militarized landscapes are not simply natural or politically neutral processes, but remain deeply bound to cultural values, politics, and broader geographies of meaning. In other words, what we do with militarized places matters and can have lasting effects on how we come to engage with and understand these places—and the policies that influence them—over the long term.

Avoiding Erasure

Globally important sites such as Hiroshima likely will always be remembered as militarized landscapes, though the particular framing of what this means and how it should be interpreted will continue to face challenge. At domestic military installations that close and transition to new types of land use, however, the prospect of losing the memory of what happened in these places and what these actions promoted in terms of our national values, politics, and environment are at much greater risk of disappearing. Military-to-wildlife conversions work doubly to naturalize sites of military production: the places become known publicly as wildlife refuges, which in turn are supposed to be *natural* places constituted largely outside the realm of culture.[10]

There are, conversely, ways that M2W conversions could serve to secure our cultural memory of the institutions and actions that predominated in creating the landscapes we now identify as new wildlife refuges. As I have already suggested, the often-dramatic hybrid qualities of these places can spur us to think more integrally about nature and society not separately, but as linked coproducers of these sites and the changes occurring here. It may be, however, that labeling these militarized/naturalized sites as "wildlife refuges" fails to fully capture this sense. It is certainly possible to be lulled by this nomenclature into an oversimplified understanding of these landscapes and how they have been created. This very concern emerged from the hearings held for the conversion of Rocky Mountain Arsenal into a wildlife refuge, as a representative of the Wilderness Society suggested that the site should not be called a national wildlife refuge because it would weaken the popular understanding of this system of lands.[11]

There is a certain irony to the Wilderness Society's expression of concern here, as the concept of wilderness itself has come under attack for its possible contributions to cultural erasure and a nature-society dualism. Environmental historian William Cronon's influential critique "The Trouble with Wilderness" highlights broader risks associated with land preservation efforts that seem to close off spaces as natural at the exclusion of the social.[12] As Cronon puts it, "Wilderness leaves precisely nowhere for human beings actually to make their living from the land."[13]

The prospective forfeiture of any lasting sense of the complex social relations built into M2W refuges stands out as one of the fundamental risks found with these conversions. As I have tried to emphasize already, this need not be the case. Redesignating former military installations to new purposes of conservation doesn't *require* that we collectively forget what happened in these places at various stages in the past, but we face this risk if we do little more than change the names of these sites and focus exclusively on the task of restoring them ecologically. Working to identify and curate these places in their greater complexity—as historical, social, and political landscapes that are also, in a sense, natural (or naturalizing)—can help prevent this loss of meaning.

Bravo 20 Environmental Memorial

The artist Richard Misrach offers at least one way through this pitfall of lost sociopolitical memory with his provocative proposal for a Bravo 20 National Park.[14] To most viewers, the Bravo 20 site scarcely conjures up associations with America's scenic national parks. This Nevada bombing range exists as a stark landscape littered with craters, bomb casings, unexploded ordnance, and the charred remains of military targets, including school buses and communications towers. In order to retain and commemorate the blend of social and environmental attributes extant in places such as the Bravo 20 bombing range, Misrach envisions an environmental memorial that invites visitors to explore, confront, and consider the site as it combines violence, power, politics, and nature:

> Bravo 20 would be a unique and powerful addition to our current park system. In these times of extraordinary environmental concern, it would serve as a permanent reminder of how military, government, corporate, and individual practices can harm the earth. . . . It would be a national acknowledgment of a complex and disturbing period in our history. . . . Bravo 20 would not only provide a graphic record of our treatment of less celebrated landscapes but also help deter their destruction in the future.[15]

Misrach's vision is explicitly political and critical—he suggests that the Bravo 20 visitor center "be devoted to the history of military abuse in peacetime. Displays and exhibits will include our radioactive experiments on the residents of the Marshall Islands in the Pacific, the contamination of continental America by tests at the Nevada Nuclear Test Site, the Colorado Rocky Flats nuclear weapons plant and Hanford nuclear area in Washington State, chemical weapons storage, toxic waste disposal, and the confiscation of land and airspace throughout America."[16] Yet the photographs and text that accompany his proposal make clear that there is also an element of beauty in this place that contributes to the project serving as a memorial for abused lands as well as a form of environmental protection. Misrach's depiction illustrates how an environmental memorial could preserve not only physical features of the land, but a sense of the processes and institutions that created these landforms. It would, in other words, maintain the visibility of the landscape's military production and press the public to learn from these actions.

Misrach is neither alone in his vision nor, in some respects, outlandish. In addition to the scenic national parks for which it is best known, the US National Park Service currently manages dozens of historical sites that recognize and commemorate military battles, massacres, and other events that do not reflect favorably upon the United States' national heritage. These include sites such as Sand Creek Massacre National Historic Site, Trail of Tears National Historic Trail, and Big Hole National Battlefield. Some national wildlife refuge managers actually gesture toward this purpose for the NPS in admiring tones. As one official at the Aroostook refuge in northern Maine told me, this M2W site would have made an excellent national park in order to preserve the place's cultural legacy. After all, the site—as Loring Air Force Base—was America's first Strategic Air Command base and the nuclear-armed US base nearest to Moscow throughout the Cold War.[17]

This points to one of the most common explanations of the value of learning about history: that we may learn from the past to inform the present and future.[18] Geographer Ken Foote addresses this concern as well, in his work on how landscapes of violence are commemorated or obliterated. When he turns in particular to how lands associated with late twentieth-century militarization are being expunged from public view, he muses, "Perhaps it would be better if more of these reminders of the Cold War were kept to commemorate a period when the entire world seemed at all times only moments away from nuclear destruction. It is my hope that these largely forgotten sites of the past fifty years will one day be marked in the landscape as reminders—and warnings—for future generations."[19]

A pivotal turn in this process of commemoration versus historical erasure comes with the Department of Defense's ability to convince the public that it has already moved beyond the problems of the past. Put more in terms of Beck's theory of risk society, the DOD seeks to generate public trust by appearing to have insured society against historical national security risks.[20] By greening military bases and committing them to new projects of environmental conservation, the federal government further works to assure the public that the military practices that produced technologies ranging from chemical weapons to nuclear bombs can be successfully managed, and that these hazards can become ecologically benign or even *helpful*. This turn, which in a very real sense naturalizes weapons,[21] strongly resembles the renegotiation of meaning that the United States pushed for in establishing the Hiroshima Peace Memorial. Militarization and military weapons, even those that bring the most horrific kinds of devastation, are no longer the problem in this view; they are the solution.

Orford Ness National Nature Reserve

On the east coast of England, Orford Ness sits as a rather nondescript stretch of shoreline, strewn with flint splinters cast inland by North Sea waves. Technically, this bit of land qualifies as the largest vegetated shingle spit in Europe, but its real claim to fame is the fascinating blend of nature and culture that has been explicitly preserved—or, more aptly, allowed to erode and evolve—since the early 1990s. Managed and owned since 1993 by the National Trust, a UK-registered charity that works to protect cultural heritage and open spaces in England, Orford Ness remains a site of surprising contrasts.

The National Trust touts Orford Ness as an "internationally important coastal nature reserve, with a fascinating military history."[22] From their first encounter, visitors to this place can't help but notice—and then be constantly reminded—that Orford Ness is not simply a reserve for yellow horned poppy, sea campion, and sea kale, or the abundant brown hares on the shingle banks and marsh harriers overhead, but it also served as an important historical site where for decades some of the United Kingdom's most advanced weapons systems, from machine guns to hydrogen bombs, were developed and tested (the fully functioning nuclear weapons were tested off-site, primarily in the Indian Ocean and Australia, but key components and ballistics were tested here).

Some of the work that was conducted at Orford Ness remains classified, but by the early 1950s the site was definitely used for early ballistics tests that proved instrumental in the British development of atomic weapons. In 1955, these operations were formalized with the creation of the British Atomic Weapons Research Establishment (AWRE) at Orford Ness; AWRE continued research on the spit until October 1971. During two decades of operations, AWRE and its predecessors conducted a variety of tests that helped solidify the United Kingdom's nuclear arsenal, particularly with research that honed weapons systems' tolerance of temperature extremes, vibration, and compatibility with a variety of delivery systems, such as submarines, aircraft, and missiles.[23]

In its curation and management of Orford Ness, the National Trust works to keep both the ecological values of the site and the relics of its military history clearly in view. As one of its tourist brochures beckons: "An international treasure of strange and scarce wildlife, littered with evidence of a frightening recent past, come to Orford Ness to feed your senses and challenge your perceptions."[24] Guided walks and tours provided here include a "Bombs & Beasties" event, and elsewhere the brochure invites tourists to "come face to face with a nuclear bomb at the first atomic weapons site open to the public anywhere in Britain."

Visitors to the Orford Ness recount "decaying forms of twisted metal and concrete [that] lure you away" from the designated trail.[25] Visitors are required to keep to the trail, lest they stray into UXO that still contaminates the site, but elsewhere are invited to explore a number of the abandoned buildings that carry their own hazards of structural collapse, jutting metal, loose cables, and other symptoms of decay (fig. 7.7). This confrontation with neglect and ruination is part of the plan designed by the National Trust, to offer visitors experiences that are not overly constrained or varnished. As one sign at the site explains: the abandoned laboratories become more evocative as they fall to ruin.[26]

Ruination, in the case of Orford Ness, also merges with processes of naturalization. Unlike the more abrupt and orchestrated naturalization of M2W refuges in the United States, at Orford Ness the ascendance of conservation goals blends more thoroughly with the gradual decline of military

7.7. Orford Ness ruins

buildings and infrastructure.[27] At least for now, both remain in view. As a result, visitors are pressed to grapple more directly with the uncertain relationships between past and present, militarization and conservation, nature and culture, and what exactly this place represents as a national reserve.

This approach that mixes curation with dereliction can be startling in many ways, so much so that the National Trust makes a point of warning visitors that they should expect to be disturbed, troubled, or challenged by what they find when they visit. Again, this differs dramatically from how the US Fish and Wildlife Service prepares visitors coming to a M2W refuge. In the "Plan Your Visit" section of the Caddo Lake refuge website, for example, two canoeists are shown paddling across a placid lake through a patch of lily pads. The website includes a passing mention of a former armory guard station that has been restored (though nothing about contamination, weapons production, or other military uses that once dominated the site); it highlights the "Wetlands of International Significance" that exist at the refuge, as well as the fact that recreational opportunities are available: "Free of charge, seven days a week from sunrise to sunset, you can enjoy wildlife-related activities, including wildlife watching, hiking, biking, hunting, and wildlife photography. Enjoy your public lands!"[28] If visitors to Orford Ness are promised something akin to a mix between a nature reserve and a Halloween spook house, those planning to head to Caddo Lake and similar US refuges are led to expect a cross between a wildlife refuge and an adventure park.

This comparison is perhaps overly glib, but I have yet to find a US refuge that works its obsolete military residues as explicitly into a public interface and the fabric of the site's meaning as Orford Ness. Even along the Iron Curtain borderlands, which I highlighted earlier for the European Union's active efforts at commemoration, the strongest initial impulse by most individuals and states was to remove the barriers and fortifications that had long characterized the militarized separation of Central Europe. In many cases, including the iconic Berlin Wall, the call to preserve standing sections of the various barriers came relatively late, in time to spare only a few scraps and small stretches, mostly now standing in the form of open-air museums (fig. 7.8).

I clearly recall the exuberance of Berliners in 1989 as they sledgehammered and chiseled the wall to pieces, rushing to rid themselves of a barrier that had marked death and division for decades. While I understand that impulse—and trust that I would have jumped at the chance to take a crack or two at the wall myself—I'm also mighty grateful that some hands were stayed. I visited a divided Berlin in 1985, and didn't make it back again until

7.8. Berlin Wall memorial

2013. Although I'd spent a full day "behind" the wall in East Berlin during my first trip, the city has changed so much since that time I was continually disoriented during my later visit, wondering which side I now was on, and how such a lethal, feared barrier could disappear so fully from view.

Of course, the wall isn't entirely gone—a Berlin Wall Trail (also the product of concerted efforts by Michael Cramer of Iron Curtain Trail credit) now outlines the former placement of the wall, and a handful of sections remain intact or reinstalled as public memorials. Berliners should be commended for trying to keep this history in view, to retain a physical memory of the past and what it represents and means, but even this remains contested and uncertain. How else to explain the renewed emphasis on building barriers across much of Central Europe (and the US-Mexico border) designed to keep certain kinds of people out? Memorialization is tenuous in the face of politics and social change, making it ever important to try to challenge, interpret, and translate the meaning of militarized places.

A Wishful Agency

One ongoing project that focuses on militarized landscapes in many of these very ways is the National Toxic Land/Labor Conservation Service, or National TLC Service. Created through "fanciful legislation" in 2011 by artist Sarah Kanouse and geographer Shiloh Krupar, the National TLC Service exists as a "wishful agency" dedicated to American militarism and its many impacts on environmental justice, human rights, and the environment. As one of its promotional videos explains, "While fully remediating the toxic legacy of the Cold War is certainly impossible, we will make sure we remember it, address it, and learn from it so we never repeat these mistakes."[29]

Although it is entirely "wishful" in the sense that Kanouse and Krupar run their agency with no formal sanction from or affiliation with the US government, the National TLC Service is patterned meticulously after official federal land management agencies such as the FWS or the NPS. "Field agents" representing the National TLC Service often sport white Tyvek suits or vests with the logo of the service fixed on a federal-style badge. A careful look at the badge shows cupped hands holding a human figure next to a mutant tree, with a radiation hazard symbol in the background against the silhouette of a flying goose—this latter reminiscent of the FWS's own cherished blue goose symbol. The badge and many of the agency's materials list an (imagined) affiliation with the Department of the Interior, a point that often confuses newcomers to the wishful agency into thinking it has some more official designation (fig. 7.9).

7.9. National TLC Service badge (image courtesy of Sarah Kanouse)

This, of course, is part of the point of the National TLC Service: to provoke thoughtful reflection about America's militarized landscapes, and to challenge existing notions about the role of government responsibility for the people and places impacted by military activities. Despite its wishful positioning, the agency does, in a sense, really exist, as Field Agents Kanouse, Krupar, and others travel with their National TLC Service "mobile field office" to art galleries, museums, and other venues to bring an end to "government unaccountability concerning the domestic effects of the American nuclear state."[30] Kanouse and Krupar also organize and facilitate design charrettes, where participants work to develop new understanding of America's militarized past, present, and future.

I attended one of these workshops in Colorado in March 2016, where thirty participants from around the region gathered to envision and design an Environmental Heritage Trail, replete with monuments and markers to bring Cold War landscapes more clearly into view. I was struck not only by the diverse perspectives the National TLC Service managed to bring together for this workshop, but also at the genuine effort each participant made to work toward a deeper collective understanding of what is important to remember about the Cold War and its legacy across the Rocky Mountain West.

What seemed at first like potentially disastrous differences in lived experience from workshop participants gradually came to be core features of productive conversation and memory construction. Two sisters, for instance, who grew up in the former uranium mill town of Uravan, Colorado, recounted in painful detail the failing health of their millworker relatives, and the eventual eradication of their town, which was declared a Superfund site, cleared of all residents, then in the late 1980s dismantled and entirely buried in an effort to contain corporate liability and contamination hazards. The personal and community loss conveyed by these women reverberated through the room, but so did the sense of pride of a man who spoke of his career with the Nuclear Regulatory Commission, working—as he viewed it—to ensure that the nuclear industry in the United States was kept to the highest standards possible. Many in the workshop could find quarrel with this view, and surely few of us could fully appreciate what it felt like to have our town "plowed up, shredded, and buried"; but the point was not so much for us each to view the world in the same way as it was to explore a variety of perspectives on the Cold War and raise awareness about how legacies of the Cold War continue to affect our "lives, lands, and bodies."[31] This combination of bridging differences and coming together in conversation to focus on the memory and meaning of militarized landscapes strikes me as exactly the kind of work we ought to be doing.

Rocky Flats, Colorado

For forty-two years, the production of plutonium triggers for the United States' nuclear arsenal took place just ten miles from the house where I grew up. For a child, this usually seemed a safe distance. Most days I was little troubled by my proximity to one of the atomic epicenters of Cold War geopolitics and its contributions to possible nuclear destruction (or prevention, as some suggest); but by the time I was a teenager, the work taking place at the Rocky Flats nuclear weapons plant sometimes came into view more personally.

I think back, for instance, to my high school physics course where a classmate's father visited one day. He worked at Rocky Flats and spoke to us about how science informed his job at the plant. His show-and-tell item was particularly memorable: a button of plutonium encased in leaded glass. He assured us it was perfectly safe in this form (the alpha radiation it emits travels just a few inches and is inept at penetrating most solids), but also emphasized that even a microgram absorbed inside the body—through the lungs, for example—could be enough to kill a person. Years later, when I learned he'd died of cancer, I wondered if somewhere at work he'd taken an inopportune breath.

When I think of Rocky Flats, I also think of the Buddhist monk who for years walked barefoot the sixteen miles from town to the plant and back, tapping a small drum in steady protest against the products of war. I remember larger protests too: the tipis pitched on railroad tracks, trying to block or delay shipments in and out of the plant; the thousands of people who held hands to encircle the plant in symbolic containment; the concerts and arrests and regular assurances of the plant's safety, then later an FBI raid and news of disastrous fires and plutonium releases and, finally, the decision to close the plant for good.[32] I remember these people and these events because I witnessed them as part of the landscape where I lived.

Today, however, when I pass this same site, only the railroad tracks are evident; all the buildings from the plant are gone. Adjacent to this former centerpiece of US weapons production, new subdivisions are springing into view. One of these, located immediately south of the former Rocky Flats site, is the Candelas development. The Candelas website presents images of verdant open space backed by the Colorado Front Range and the tag line "Life Wide Open is Our Dream View." The accompanying text sells the environmental amenities of the development: "There is a magnificent sweep of mountain pastureland you'd swear you've seen before on picture postcards of the great American West. This wide-open landscape, this epitome of raw western beauty, is called Candelas. . . . Candelas presents a life full of the very things people love most about Colorado. Come live life wide open."[33] The development also has a website dedicated specifically to the Rocky Flats NWR, where prospective homeowners can easily find more images of wildlife, open space, and mountain scenery amid assurances of the location's safety, but readers need to dig deep to find a single mention of plutonium.[34]

My point here is not to claim that the Candelas development or even the Rocky Flats site itself is not safe—I hope that they are. Rather, I want to raise questions about what it is that most people now come to know about this place. Presented with a landscape that no longer appears militarized, where the protests have quieted and production has long since stopped, should we—or those buying homes at Candelas—accept this land as natural, the epitome of raw western beauty, or is it important to think about it differently because of its past?

This happens to be a place that means something to me, a place whose history to me will *always* seem notable in a cautionary way, but I am far from alone in questioning how Rocky Flats now is represented by developers and others keen to move past the site's forbidding history. In an effort to raise awareness in prospective homeowners, activists from the downwind community of Arvada, Colorado, and surrounding areas have created alternative Candelas websites. The Candelasconcerns.com site seeks "to make sure [new home buyers interested in Candelas] are aware of possible contamination from plutonium and other materials." The site also notes that there is no requirement to inform prospective buyers about the history of the former weapons site or any risks that may come from living adjacent to it.

Another alternative website, Candelasglows.com, highlights some of the many problems of burying the past at Rocky Flats:

> Candelas, one of Colorado's largest new suburban developments, is part of an alarming trend of forgetting about its neighbor, Rocky Flats—a former Nuclear Weapons Plant and Superfund site. . . . We believe Rocky Flats needs to be remembered for what it is, with plant workers recognized as the veterans they are. The "wildlife refuge" designation needs to be immediately stripped and *not* opened to the public. We believe the site should be memorialized, calling on artists to help us build permanent structures that speak to the site's past much the way other historical tragedies are memorialized. A memorial could also commemorate the workers and neighbors who have been deeply impacted by the legacy of the site.

Hanford, Washington

I would like to think that my experiences with militarized landscapes are somehow extraordinary, that I encounter these places only because I seek them out, but I know this isn't entirely true. My experiences *are* different; the contexts and contours of my interactions are unique. Most of us, though, have encounters like these in some form or another. Perhaps yours is the Hanford site, in Washington State, which now draws tourists to visit the B Reactor that produced plutonium used in the Trinity test in New Mexico, and the bomb that devastated Nagasaki. The reactor is part of Manhattan Project National Historical Park, which was designated in November 2015 and includes places that were instrumental in America's early development of atomic bombs: Hanford, Washington; Oak Ridge, Tennessee; and Los Alamos, New Mexico. The park is unique for its dispersed geography, its joint management by the NPS and the US Department of Energy, and also highlights how commemoration and interpretation of militarized landscapes can be accomplished in more nuanced ways than many military-to-wildlife refuges seem to manage. As the NPS describes this new historical park:

> The Manhattan Project and its legacy is a complex story. It's the story of more than 600,000 Americans leaving their homes and families to work on a project they were told was vital to the war effort. It's the story of generals, physicists, chemists, mathematicians, and engineers pushing and broadening the limits of human knowledge and technological achievement in ways never before imagined. It is also the story of the death and destruction associated with World War II and a new weapon capable of unimagined levels of devastation. A visit to the Manhattan Project National Historical Park . . . challenges us to think about how the world has changed with the dawn of the nuclear age.[35]

The Hanford portion of the Manhattan Project park is also contained within a larger protected area managed by the FWS: the Hanford Reach National Monument. As yet another form of an M2W landscape, Hanford Reach is acclaimed as one of the "largest river complexes in the country" that holds "an exceptionally wide variety of habitats within a relatively small assemblage of public lands."[36] In his June 2000 declaration that created the monument, President Bill Clinton described Hanford Reach as "a biological treasure," created in part by its location as a buffer surrounding the nuclear weapons development that took place.[37] The FWS points to the site's desert and river habitats as "sharply contrasting environments," a description that could also fit the ecological prominence of Hanford Reach in contrast to its legacy of environmental contamination. The reach itself, a fifty-mile segment of the Columbia River, is considered the longest free-flowing stretch of the western United States' largest river, and has been proposed repeatedly for national Wild and Scenic River designation. (Geographer Shannon Cram puts the unnatural history of the reach more explicitly in view, in one article calling Hanford a "Wild and Scenic Wasteland.")[38]

US Senator Patty Murray (D-Washington) was among those who lauded the monument designation. In her press release, delivered in 2000 as she prepared to float a section of the river with Vice President Al Gore, Murray framed the new monument primarily as an extraordinary natural landscape: "From its pristine natural beauty to the salmon who spawn in its waters to its strong Native American history, the Hanford Reach is a unique American resource. This designation means more salmon restoration, more recreation and tourism, and national prominence for the Tri-Cities community and Washington State."[39] Though Murray alluded obliquely to a "community that has given so much to . . . our country," no doubt with the Tri-Cities of Kennewick, Pasco, and Richland, and Washington State's history of weapons production in mind, she failed to include a single mention of nuclear weapons production, contamination, or the heavy traces of industrial civilization that in so many ways shape this site.

In fact, it took years of concerted effort to ensure that all remaining infrastructure from the various Manhattan Project sites wasn't simply obliterated. By the early 1990s, with Cold War production facilities closing and facing costly cleanups, many officials deemed razing and burial to be the most cost effective and attractive option for dealing with obsolete militarized landscapes. This approach was implemented in places such as Rocky Flats, the Rocky Mountain Arsenal, and many of the BRAC closures that have since become wildlife refuges. At the Los Alamos National Laboratory, officials estimated it would cost $3 million simply to stabilize aging buildings. In 1997, the director of the lab declared that preserving the buildings "would be a waste of taxpayers' money."[40]

Against these plans to demolish key properties at Los Alamos and similar proposals to dismantle the Hanford reactors, a handful of voices spoke against the loss of key elements of American Cold War history. One Department of Energy employee, Cynthia Kelly, contacted the federal Advisory Council on Historic Preservation and urged they visit the Los Alamos site. In November 1998, members of the advisory council met at Los Alamos and concluded not only that the facility ought to be protected as a national historic landmark, but that it was suitable for designation as a World Heritage Site (Japan's Hiroshima Peace Memorial received this designation in 1996). Kelly subsequently stepped away from a decades-long career with the Department of Energy to found and direct the nonprofit Atomic Heritage Foundation, which ultimately played a key role in securing lasting federal recognition and protection of Manhattan Project sites.[41]

The national historical park designation that now covers Los Alamos, Oak Ridge, and Hanford helps establish that important cultural elements should not be lost from these sites, even as Hanford's national monument designation marks its ecological features. The two kinds of protections applied to Hanford, at least, may serve to prevent losses of meaning that narrower wildlife refuge labels seem to accommodate. At the very least, the treatment of Hanford's complex mix of land uses and conditions illustrates how federal agencies and elected officials can respond more effectively to preserve nature and culture together. It may not be easy—as the Atomic Heritage Foundation's Cynthia Kelly points out, it took longer to create Manhattan Project National Historical Park than it did for the actual Manhattan Project to develop an atomic bomb—but this and other examples show how it can be done.

Legacies of Militarization

Encounters with militarized landscapes come in diverse form, many of which may be unexpected, unplanned, or unknown. Of the two hundred thousand annual Washington, DC–area visitors to the Patuxent Research Refuge, I wonder how many realize that the FWS sometimes has to close its lands on short notice to accommodate military training activities and aerial drop zones. Visitors to Patuxent's North Tract still have to sign a liability waiver due to the lingering presence of UXO in many areas, but how many are told—as I was, only when I interviewed officials there—that a sweep of selected areas in 2003 and 2004 turned up more than ten thousand munition items?

Or what of the birders at Occoquan Bay NWR on the Virginia coast? The abandoned telecommunications posts at Occoquan now support osprey nests rather than tests of electromagnetic pulses directed at armored vehicles, but both the birds and the blasts of energy surely shape the meaning of this place, just as diving amid the wrecks at Bikini Atoll or bicycling the borderlands of the Iron Curtain remain linked to the dramatic histories of those sites.

Many of us remember these places because in some way they have become personal for us. They are sites where we've worked or visited, sites we recognize for their contributions to national defense and security, or sites we've avoided because of their hazards and contamination. In many cases, these are not just natural or naturalizing spaces once possessed or impacted by the military, but places that can press us to think about how the natural and the cultural fit together. Even as it remains impossible to remember everything about each of these sites, it also remains important to resist forgetting too much about certain kinds of places. This recognition is precisely what spurred the designation of the Manhattan Project Historical Park in the United States and the Iron Curtain Trail across Central Europe. There is, after all, a cost to losing sight even of landscapes we might like to avoid; there is a cost to forgetting. As Viet Thanh Nguyen points out, "We forget despite our best efforts, and we also forget because powerful interests often actively suppress memory. . . . Nations cultivate and would monopolize, if they could, both memory and forgetting."[42]

I often think back to the days I've spent in these militarized but also naturalizing landscapes, and sometimes I wonder what I too will remember. After all, I'm drawn to many of these places not so much because of their militarization, but because they are changing. The Rocky Mountain Arsenal fascinates me every time I visit, but most days what stirs me are the bison and the prairie dogs, the coyotes and the eagles, the open prairie and the chance of seeing a burrowing owl. In those moments, those glimpses of an American prairie I can scarcely imagine, it's easy to lose sight of the Tokyo neighborhoods and Vietnamese farms and forests charred by the incendiaries that were produced here, the lethal menace of sarin gas, the nearby wells contaminated by chemicals manufactured with a singular focus on national defense, and the lives and livelihoods torn apart by what happened here. Each of these characteristics—the beauty and the horror, the promise of a better future and the damage caused in the past—shapes the meaning of this place. I understand the mixed heritage of militarization, conservation, and ecological restoration that exists here, but I also see how quickly meaning can fade from a landscape. I appreciate this and similar places more fully by thinking carefully about where they have been, how we made them, and what they have yet to teach us.

This, then, is both the promise and the peril of how we interact with these landscapes. The lessons and legacies these places provide remain shaped not just by the dynamic processes of ecosystems and physical change, but also by politics, agency budgets, and social priorities. How we allow ourselves to experience and interpret these sites will shape their meaning as we encounter the past. How this translates to policy and action will, in turn, affect the way we live today and in the future.

ACKNOWLEDGMENTS

The genesis for this book was a brief conversation, nearly two decades ago, that prompted me to think about the relationship between militarized landscapes, ecological restoration, and conservation. My inquiries, explorations, and ideas about these complex landscapes have branched in many directions since that time, but my curiosity about these places hasn't waned. This is in part due to the variability of these sites, many of which have borne staggering impacts from military training, testing, or conflict, yet now attract attention as hotspots of biodiversity more than contamination. It's also a reflection of the many people I've talked to along the way, each with their own connections to similar kinds of places, whether as land managers, scholars, visitors, military personnel, or nearby residents. This book is a reflection of many of those interactions. Any flaws, I'm afraid, are my own contribution.

Some of the material in this book draws upon previously published articles. These include "The Iron Curtain Trail's Landscapes of Memory, Meaning, and Recovery," *Focus on Geography* 57, no. 2 (2014): 126–33; "Examining Restoration Goals at a Former Military Site," *Nature and Culture* 9, no. 3 (2014): 288–315 (coauthored with Marion Hourdequin and Matthew John); "Opportunistic Conservation at Former Military Sites in the United States," *Progress in Physical Geography* 38 (2014): 271–85; "Restoration and Authenticity Revisited," *Environmental Ethics* 35, no. 1 (2013): 79–93 (coauthored with Marion Hourdequin); "Disarming Nature: Converting Military Lands to Wildlife Conservation," *Geographical Review* 101, no. 2 (2011): 183–200; "Ecological Restoration in Context: Ethics and the Naturalization of Former Military Lands," *Ethics, Policy, and Environment* 14, no. 1 (2011): 69–89 (coauthored with Marion Hourdequin); "Logics of Change for Military-to-Wildlife Conversions in the United States," *GeoJournal* 69 (2007): 151–64.

I'm very grateful to Christie Henry, at the University of Chicago Press, for her steady hand, encouragement, and support for this project at every turn. She's been a joy to work with. Miranda Martin, Christine Schwab, and Nick Lilly at the press kept everything on track during production, and Johanna Rosenbohm's copyedits made this a cleaner, better book. I also very much appreciate the reviews of earlier versions of this proposal and manuscript. Comments from Kate McCaffrey, Bethanie Walder, and four anonymous readers were thoughtful, critical, and constructive.

My early research on the topic of military-to-wildlife land use changes was supported by a number of people and sources at the University of North Carolina (UNC) at Chapel Hill. Scott Kirsch advised me with candor, keen insight, and the occasional milkshake or beer. The Geography Department at UNC was a wonderful early academic home. Tom Whitmore, Wendy Wolford, Larry Band, Steve Walsh, Leo Zonn, Wil Gesler, Altha Cravey, Chip Konrad, John Florin, Aaron Moody, Rebecca Vidra, and Banu Gökariksel each mentored me in particular ways. John Pickles very generously included me in a number of his seminars and advising groups. Martin Doyle helped me think ecologically and has been a friendly spur in my side over time, each year asking, "When are you going to write the book?"

I also had a terrific batch of colleagues—thanks to Tim Baird, Dilys Bowman, Brian Doyle, Clark Gray, Jon Lepofsky, Tina Mangieri, Carlos Mena, Christian Sellar, and Dan Weiss. Outside the Geography Department, Robert Cox, in Communication Studies at UNC; Lisa Campbell, at Duke's Nicholas School of the Environment; and the late John Richards, at Duke, made time for me whenever I came asking and pressed me to consider new perspectives on environmental communications, conservation narratives, and environmental history, respectively.

I am grateful to the National Science Foundation's Science and Technology Studies for a Doctoral Dissertation Research Improvement grant (number 0521728) and to the UNC Department of Geography's Eyre Travel Fund, which supported my early research.

I have turned to an array of federal, state, and county officials, librarians, and citizen activists throughout my research and writing to glean information and insights about the military conversions at the core of my project. Refuge officials from the US Fish and Wildlife Service provided important on-the-ground perspectives and granted me access to refuge lands, archival documents, and their considerable knowledge and experience. These public employees deserve far more public recognition and financial support than they get. I benefited repeatedly from the time, generosity, and expertise shared by Steve Agius, Alan Anderson, Dionne Briggs, Paul Bruckwicki,

Dwight Cooley, Sherry James, David Jones, Brad Knudsen, Bill Kolodnicki, Bob Leffel, Jason Lewis, Dan Matiatos, Steve Miller, Nancy Morrissey, Susan Rice, Jason Roesner, Tom Ronning, Dean Rundle, Mark Sattelberg, Bill Stephenson, Melissa van Dreese, Greg Weiler, Brian Winters, and Terry Wright.

John Davis and Bruce Hastings hosted my visits, with and without students, at the Rocky Mountain Arsenal year after year, and they have been more helpful than I ever could have hoped. Libby Herland, Joe Robb, and Graham Taylor welcomed me to their refuges for multiple visits and interviews and patiently fielded questions, phone calls, and tours without fail. Many other FWS employees, volunteers, Environmental Protection Agency personnel, Department of Defense officials and contractors, and representatives from citizen groups were generous with their time and knowledge during fieldwork I conducted at refuges and former military sites. Thank you in particular to Ken Knouf and Richard Hill in Madison, Indiana, for meeting with me on multiple occasions; and to Paul Boothroyd and Jan Wright, who brought the bunkers to life at the Assabet River refuge.

At times throughout the book I draw upon information or quote from interviews conducted with people listed here, and others. Although very few of these participants insisted on anonymity, wherever possible I have tried to protect their privacy and maintain a degree of confidentiality.

I'm grateful to the National Science Foundation for a multiyear grant (award number 0957002, with co-PI Marion Hourdequin) that provided essential support for the project from which this book eventually emerged. As part of this grant, I was fortunate to work with several talented student research assistants at various stages of the project. My thanks to Reginald Anderson, Jon Harmon, Matthew John, and Claire McCusker. A number of the ideas included in the book benefited from conversations and presentations offered during an NSF-supported workshop in 2013 that Marion Hourdequin and I organized to focus on the restoration of layered landscapes. In addition to FWS officials already named, other participants who shared ideas and insights there include Andre Clewell, Justin Donhauser, Martin Drenthen, Robert Earle, Erica Elliott, Rebecca Garvoille, William Jordan III, Jozef Keulartz, Jennifer Ohayon, and Allen Thompson.

I received funding from the American Geographic Society as a McColl Family Fellow to support my travel to Central Europe for research along the Iron Curtain Trail and related borderlands. Michael Cramer, Member of Parliament (EU), generously took time out of his day to meet me for coffee, hot chocolate, and conversation in Berlin. His efforts to commemorate the history, culture, politics, and nature of the Berlin Wall and Iron Curtain—and to get everywhere by bicycle—continue to inspire me. My research in

Europe was also supported by a grant from the Committee for Research and Creative Work at the University of Colorado Colorado Springs (UCCS).

My research in Japan was made possible by support from the Global Intercultural Research Center at UCCS. I very much appreciate the funding. I remain fortunate to have wonderful colleagues in the Department of Geography and Environmental Studies at UCCS. On multiple occasions they read early manuscripts of articles that grew, incrementally, into this larger project. Thank you to Eric Billmeyer, Diep Dao, Somayeh Dodge, Cerian Gibbes, John Harner, Paddington Hodza, Curt Holder, Carole Huber, Tom Huber, Steve Jennings, Irina Kopteva, Bob Larkin, Mike Larkin, Emily Skop, Rebecca Theobald, and Brandon Vogt. Nearby at Colorado College, Eric Perramond has proved to be an excellent resource and collaborator. I've also benefited from conversations or correspondence with a number of scholars farther afield, including Alec Brownlow, Peter Coates, Tim Cole, Caitlin DeSilvey, Marianna Dudley, Ryan Edgington, Matthew Farish, Matthias Gross, Jamie Lorimer, Harlan Morehouse, Chris Pearson, Harold Perkins, Sonja Pieck, Heather Swanson, Julia Adeney Thomas, Richard Tucker, and Rachel Woodward.

My research collaborators for trips to militarized landscapes often came from closer to home: Adele and Tim Havlick have become two of my most energetic and delightful research companions. What fine good fortune to have them. My parents, Spense and Val Havlick, have supported me in more ways than I can even list. From joining me on research trips to providing childcare so I could slip away for a few hours or days, they've been stalwart. My sister Jenny and nephew Spenser enthusiastically joined me for a day touring Rocky Mountain Arsenal and looking for burrowing owls. My brother, Scott, remains my faithful legal counsel and reality check. Mary and Jim Hourdequin helped launch me to New England–area refuges and also provided childcare, thoughtful questions, and swims in the lake to keep me going. Peter and Junko Hourdequin helped motivate, facilitate, and translate my visits to Japan. Jim and Kathryn Hourdequin brought me as close as I could get to Nomans Land Island and fit in some fine days camping along the way.

The biggest share of thanks goes to Marion Hourdequin. To call her my most reliable collaborator falls far short, though it's also true. She first alerted me to the fire management plan for the Big Oaks National Wildlife Refuge, and her curiosity and knowledge of ecology, philosophy, restoration, and conservation continue to make me think more carefully about these places and how they—and we—fit in the world. This book wouldn't exist without her.

NOTES

CHAPTER ONE

1. A *spotting charge* is a small explosive designed to set off a flash and/or smoke pulse from a military projectile to help identify impact locations on test ranges.
2. European Green Belt Initiative, "Borders Separate. Nature Unites!"
3. See Latour, *We Have Never Been Modern*.
4. See, for example, McKibben, *The End of Nature*; Whatmore, *Hybrid Geographies*; Castree, "Commentary." The idea of human domination is also a central tenet of the new epochal designation of the Anthropocene. See, for example, Steffan, Crutzen, and McNeill, "The Anthropocene"; and Zalasiewicz et al., "The Anthropocene: A New Epoch?"
5. In June 2014, President Barack Obama announced plans to expand the Pacific Remote Islands Marine National Monument, which includes a number of former military installations and test sites. While the designation differs in name from a wildlife refuge, its management and intent is much the same and, when completed, will include wildlife refuges as individual units and expand the area of M2W conversions dramatically by placing more than 500 million acres into protected status. See Eilperin, "Obama Proposes Vast Expansion of Sanctuaries."
6. Woodward makes a similar point in *Military Geographies*.
7. See, for example, Coates, "Borderland, No-Man's Land, Nature's Wonderland."
8. Details on the history of the site come from US Fish and Wildlife Service, Caddo Lake, "History of the Longhorn Army Ammunition Plant," and interviews I conducted with refuge officials on July 11, 2006 (by phone), and March 23, 2011 (in person).
9. US Fish and Wildlife Service, Caddo Lake, "Wildlife & Habitat" and "Paddlefish Restoration and Recovery."
10. US Fish and Wildlife Service, "Rocky Mountain Arsenal National Wildlife Refuge," 5.
11. Details on the development interests come primarily from author interviews conducted with refuge personnel and long-term residents of nearby Marshall, Texas, July 11, 2006 (by phone), and March 23, 2011 (in person).
12. Hancock, "Don Henley Works to Preserve Caddo Lake"; see also "Dwight Killian Shellman, Jr.," obituary, accessed June 15, 2016, http://www.caddolakeinstitute.us/docs/DKS%20Obit%2020120326.pdf; and Caddo Lake Institute, "Our Mission."
13. Quoted in Hancock, "Don Henley Works to Preserve Caddo Lake."

14. See Finley, "Security Water Raises Concern."
15. "From Swords to Swards" comes from a conference presentation shared with me by a refuge worker at Rocky Mountain Arsenal in 2005; "Bombs to Birds" is from the title of a documentary film made about Caddo Lake National Wildlife Refuge in 2012 by Richard Michael Pruitt.
16. See Elliot, "Faking Nature" and *Faking Nature*.
17. See Hourdequin and Havlick, "Restoration and Authenticity Revisited."
18. Geidezis and Kreutz, "Green Belt Europe."
19. See Harvey, *Justice, Nature and the Geography of Difference*; Christoff, "Ecological Modernisation, Ecological Modernities," 496; and Hajer, "Ecological Modernisation as Cultural Politics."

CHAPTER TWO

1. Dumanoski, "Pentagon Takes Steps toward Tackling Pollution."
2. See, for example, Palka and Galgano, *The Scope of Military Geography*; and more critically, Farish, *The Contours of America's Cold War*.
3. Office of the Undersecretary of Defense, "Operation and Maintenance Overview."
4. Benton, Ripley, and Powledge, *Conserving Biodiversity on Military Lands*, 21.
5. See US Department of the Army, "Sustainability Report 2014."
6. US Fish and Wildlife Service, "Budget Justifications and Performance Information, FY 2015," EX-3.
7. US Fish and Wildlife Service, "Budget Justifications and Performance Information, FY 2015," EX-3.
8. A "major installation" in the United States contains at least 10 acres and $1.5 million in assets, according to the US Department of Defense, "Base Structure Report: Fiscal Year 2003 Baseline." On the total number of sites affected by base closure, see US Department of Defense, "Base Structure Report: Fiscal Year 2003 Baseline"; and 2005 Defense Base Closure and Realignment Commission, final report, p. iii.
9. Interview with Barbara Wyman, Realty Division program manager for base conversion lands, US Fish and Wildlife Service, Washington, DC, October 6, 2003; interview with D. Vandegraft, chief cartographer, Realty Division, US Fish and Wildlife Service, October 16, 2003; interview with Linda Shaffer, chief of Cartography and Spatial Data Services Branch, US Fish and Wildlife Service, Hadley, MA, January 19, 2004; interview with Cathy Osugi, BRAC coordinator, Division of Refuge Planning, US Fish and Wildlife Service, Portland, OR, February 25, 2004. At some of these sites, land title remains with the DOD. Acreages include several Pacific Island sites that contain extensive marine holdings.
10. The Alameda National Wildlife Refuge (formerly Naval Air Station Alameda) in the San Francisco Bay area has already been identified as an active M2W site—the FWS requested 900 acres for use as a wildlife refuge—but transfer was stalled by questions relating to cleanup and long-term liability. In 2013, 549 acres at the site were transferred to the US Department of Veterans Affairs, who will manage most of the land for conservation and open space purposes. The FWS continues to manage a population of endangered California least terns that reside at the site. See Alameda Point Info, "What Is Happening with the Refuge?"; and Alameda Point Environmental Report, "Film: 'Demilitarized Landscapes.'"
11. See Proclamation 8803—Establishment of the Fort Ord National Monument.
12. Shulman, *The Threat at Home*; Leslie, Meffe, and Hardesty, *Conserving Biodiversity on Military Lands*; as Fischman notes in *The National Wildlife Refuges*, 22: "Severe restric-

tions on public access to these [DOD] lands have preserved important wildlife habitat. On the other hand, secrecy enshrouding management of these lands has led to instances of appalling degradation and a collection of the most severely contaminated sites in the country."

13. US Fish and Wildlife Service, "Former Bombing Range Becomes Refuge." The new refuge is managed as an overlay with army ownership still in place with FWS management; this quote from Clark and the earlier remarks from Cheney are but two examples of this overlay. See also broader treatments such as Lillie and Ripley, "Implementing Ecosystem Management in the United States Air Force"; and Hoffecker, *Twenty-Seven Square Miles*.
14. Woodward makes a similar point in *Military Geographies*, 102.
15. Woodward, *Military Geographies*, 9.
16. Goren, *The Politics of Military Base Closings*, 46; Sorenson, *Shutting Down the Cold War*, 31.
17. Sorenson, *Shutting Down the Cold War*.
18. It's worth noting that over the longer term, studies have found that base closures often actually benefit local economies and prompt greater economic diversification. See, for example, US General Accountability Office, "Military Base Closures: Overview of Economic Recovery, Property Transfer, and Environmental Cleanup"; and Cowan, "Military Base Closures: Socioeconomic Impacts."
19. US Department of the Army, "Environmental Protection and Enhancement," 48.
20. See, for example, US Department of Energy, "Stewards of National Resources"; Joanna Burger, "Integrating Environmental Restoration and Ecological Restoration"; Burger et al., "Shifting Priorities at the Department of Energy's Bomb Factories."
21. See, for example, Loeb, "Unexploded Arms Require Big Cleanup."
22. See, for example, Tangley, "Bases Loaded"; and DENIX, "2012 Secretary of Defense Environmental Awards."
23. See DENIX, "2015 Secretary of Defense Environmental Awards."
24. See, for example, US Department of the Army, "Jefferson Proving Ground Final Environmental Impact Statement"; Tierney, "Case Study of Great Bay National Wildlife Refuge," Broad, "Bid to Preserve Manhattan Project Sites Stirs Debate."
25. Fischman, *National Wildlife Refuges*, 36.
26. See, for example, Olsen, "At Hanford, the Real Estate Is Hot"; and Hansen, "Free-Flowing Debate Control over Hanford Reach."
27. US Government Accounting Office, "National Wildlife Refuges: Continuing Problems," 16.
28. US House of Representatives, Committee on Natural Resources, "Buying More Land When We Can't Maintain What We Already Own." Funds made available by the 2009 American Recovery and Reinvestment Act were the major cause of this improvement.
29. For example, a 1912 executive order designated Nebraska's Fort Niobrara a refuge for migratory birds. See US Fish and Wildlife Service, Fort Niobrara, "About the Refuge."
30. US Fish and Wildlife Service, National Wildlife Refuge System, "About: Mission."
31. See Fischman, *The National Wildlife Refuges*.
32. US Fish and Wildlife Service. "Assabet River National Wildlife Refuge."
33. US Fish and Wildlife Service, Big Oaks, "About the Refuge."
34. US Department of the Army, "Jefferson Proving Ground Final Environmental Impact Statement," 4-41.
35. National Wildlife Refuge System Improvement Act of 1997, Public Law 105-57, US Congress 111 Stat. 1252, October 9, 1997.

36. Krupar highlights this vividly in *Hot Spotter's Report: Military Fables of Toxic Waste.*
37. Consent decree between the United States of America and the State of Colorado, in the US District Court for the District of Colorado, Civil Action No. 83-C-2386, State of Colorado, Plaintiff, vs. United States of America, Shell Oil Company, et al., Defendants, filed February 27, 2009. See also State of Colorado v. US Department of the Army, 707 F. Supp. 1562 (D. Colo. 1989), memorandum opinion and order, accessed July 3, 2017, https://casetext.com/case/state-of-colo-v-us-dept-of-army.
38. Allard, introduction of legislation establishing Colorado Metropolitan Wildlife Refuge.
39. US Fish and Wildlife Service, Nomans Land Island, "About the Refuge."
40. Interview with anonymous US Navy cleanup personnel, Vieques National Wildlife Refuge, January 11, 2010.
41. See Machlis and Hanson, "Warfare Ecology"; Pearson, Coates, and Cole, *Militarized Landscapes*; Dudley, *An Environmental History of the UK Defence Estate.*
42. Woodward makes this point emphatically in *Military Geographies.*
43. For discussion of some of the philosophical challenges relating to this, see Drenthen, "Ecological Restoration and Place Attachment"; and Hourdequin and Havlick, "Ecological Restoration in Context." More broadly, see Hourdequin and Havlick, *Restoring Layered Landscapes.*
44. See, for example, Guy, "'Bomblet' at Arsenal Cancels All Tours"; and Nazaryan, "Tiptoeing through America's Sarin Stash."
45. Associated Press, "Crews Find No New Signs of Chemical Weapon"; and Nazaryan, "America's Sarin Stash."
46. Society for Ecological Restoration, International Science & Policy Working Group, "Primer on Ecological Restoration."
47. Society for Ecological Restoration, International Science & Policy Working Group, "Primer on Ecological Restoration." It is important to note that not all restoration scholars subscribe to SER's version of reference conditions. See for example, Holland and O'Neill, "Yew Trees, Butterflies, Rotting Boots and Washing Lines"; Choi, "Theories for Futuristic Restoration in Changing Environment"; and Glass, Herrick, and Kucharik, "Climate Change and Ecological Restoration at the University of Wisconsin–Madison Arboretum."
48. Woodworth provides a lively overview of the breadth of activities restoration ecologists face in his book *Our Once and Future Planet: Restoring the World in the Climate Change Century.*
49. See, for example, Ripley and Leslie, "Conserving Biodiversity on Military Lands"; Armstrong, McDermott, and Ripley, "The U.S. Air Force Embraces Ecosystem Management"; Durant, *The Greening of the U.S. Military*; and Machlis and Hanson, "Warfare Ecology."
50. See Anderson and Havlick, "History and Values in Ecological Restoration Workshop."
51. Rocky Mountain Arsenal Remediation Venture Office, RVO fact sheet.
52. Tetra Tech EC, Inc., for Rocky Mountain Arsenal RVO, "Final 2010 Five-Year Review Report for Rocky Mountain Arsenal, 2010."
53. Society for Ecological Restoration, International Science & Policy Working Group, "Primer on Ecological Restoration."
54. For a general overview of the guiding principles of ecological restoration, see the Society for Ecological Restoration, "Mission and Vision."

CHAPTER THREE

1. US Fish and Wildlife Service, Rocky Mountain Arsenal, "About the Refuge."
2. This distinction between ecological restoration as a set of practices or the application of the *science* of restoration ecology is sometimes blurred, but one I will try to adhere to in this chapter. See, for example, Galatowitsch, *Ecological Restoration*.
3. Langston, "Restoration in the American National Forests," 164.
4. Langston, "Restoration in the American National Forests," 164.
5. Langston, "Restoration in the American National Forests," 164.
6. See Bradshaw, "What Do We Mean by Restoration?"
7. Langston makes a similar point in "Restoration in the American National Forests," 194.
8. See, for example, Botkin, *Discordant Harmonies*.
9. Cronon's 1995 essay "The Trouble with Wilderness" remains one of the more influential versions of this critique.
10. See, for example, Light, "Ecological Restoration and the Culture of Nature."
11. Hourdequin does a nice job working through many of these distinctions in *Environmental Ethics*.
12. Society for Ecological Restoration, International Science & Policy Working Group, "Primer on Ecological Restoration."
13. Hobbs, Higgs, and Hall, *Novel Ecosystems*; Hobbs, Higgs, and Harris, "Novel Ecosystems."
14. Choi, "Restoration Ecology to the Future" and "Theories for Ecological Restoration in Changing Environment."
15. US Fish and Wildlife Service, Rocky Mountain Arsenal, "Resource Management."
16. For more detail on the survey and its results, see Havlick, Hourdequin, and John, "Examining Restoration Goals at a Former Military Site." The survey collected responses from a total of 112 refuge visitors.
17. To be clear, SER has never fully jettisoned historical conditions as a broad guide for restoration efforts. In the same 2004 "Primer on Ecological Restoration," SER declares, "Restoration attempts to return an ecosystem to its historic trajectory. Historic conditions are therefore the ideal starting point for restoration design."
18. See Elliot, "Faking Nature." This section on authenticity in restoration was treated in a previous publication: Hourdequin and Havlick, "Restoration and Authenticity Revisited."
19. Egan, "Authentic Ecological Restoration."
20. Hourdequin and Havlick in "Restoration and Authenticity Revisited" describe this as ontological authenticity, relating to the material condition of a site, which we then contrast with epistemological authenticity, or what it is one can learn from a given site.
21. This agreement extends across county, state, and federal regulators, as well as the principal parties who have operated at the site in recent decades: the army, FWS, and Shell. Some citizen groups and individuals, such as the Site Specific Advisory Board, have continued to raise serious concerns about the legitimacy of the cleanup and its reliability. See, for example, the April 8, 2011, citizen report from the Site Specific Advisory Board to the US Army in Jackson, Lavenue, and Singh, "Review of Long-Term Monitoring Plan for Rocky Mountain Arsenal," starting at page 10.
22. Berry, "The Futility of Global Thinking."
23. See US Fish and Wildlife Service, "Vieques Comprehensive Conservation Plan and Environmental Impact Statement"; or more broadly, McCaffrey, *Military Power and Popular Protest*.

24. See US Environmental Protection Agency, "Superfund: Atlantic Fleet Weapons Training Area."
25. US Navy, Naval Facilities Engineering Command, "Former Atlantic Fleet Weapons Training Area—Vieques."
26. Bearden, "Vieques and Culebra Islands," 8–9; Fox, "U.S. Rattles Puerto Rico with Cleanup Plan."
27. Becker, "Vieques: Long March to People's Victory."
28. McCaffrey, "Fish, Wildlife, and Bombs."
29. See DENIX, "FY 2009 Secretary of Defense Environmental Award."
30. See, for example, McCaffrey, "Fish, Wildlife, and Bombs."
31. Navarro, "Navy's Vieques Training May Be Tied to Health Risks."
32. US Department of Health and Human Services, "An Evaluation of Environmental, Biological, and Health Data from Vieques."
33. See, for example, Colón-Ramos, "Letter to President Obama about Vieques."
34. "Testimony of John P. Wargo, May 20, 2010, before the Subcommittee on Investigations and Oversight, House Committee on Science and Technology, US House of Representatives," p. 2, accessed January 6, 2016, http://archives.democrats.science.house.gov/Media/file/Commdocs/hearings/2010/Oversight/20may/Wargo_Testimony.pdf; see also Wargo, *Green Intelligence*, 90–94 and passim.
35. Navarro, "New Battle on Vieques, over Navy's Cleanup of Munitions," A10.
36. Higgins, "Affordable Caribbean: Vieques."
37. US Fish and Wildlife Service, "Vieques Comprehensive Conservation Plan and Environmental Impact Statement," 1.
38. US Fish and Wildlife Service, "Vieques Comprehensive Conservation Plan and Environmental Impact Statement," 1.
39. *Alamogordo Daily News*, "Trinity Site to Be Open April 4." See also White Sands Missile Range, "Trinity Site Open House."
40. Edgington, *Range Wars*.
41. US Fish and Wildlife Service, "Vieques Comprehensive Conservation Plan and Environmental Impact Statement," 66.
42. McMenemy, "EPA Orders Air Force to Treat Wells."
43. US Fish and Wildlife Service, Great Bay, "About the Refuge."
44. US Fish and Wildlife Service, Great Bay, "About the Refuge."
45. National Park Service, Gettysburg National Military Park, "Management."
46. National Park Service, Little Bighorn Battlefield, "Little Bighorn Battlefield National Monument Resources Management Plan," 3.
47. National Park Service, Trail of Tears, "A Journey of Injustice."
48. Todd Lookingbill, email to the author, November 29, 2015.
49. National Park Service Organic Act of 1916, 16 U.S.C. 1. A full accounting of the National Park Service's ability to manage for historic sites and cultural features should also include the 1906 Antiquities Act, which authorizes the US president to protect "historic landmarks, historic and prehistoric structures, and other objects of historic or scientific interest" as national monuments (most of which are subsequently managed by the NPS), and the Historic Sites Act of 1935, which expanded the Park Service's ability to research, restore, preserve, and maintain historic properties and objects. The 1916 act, however, remains the most significant guiding legislation for the agency.
50. US Fish and Wildlife Service, National Wildlife Refuge System, "National Wildlife Refuge System Improvement Act of 1997." Environmental policy scholars make this

point about the agency's dedication to wildlife as well. See, for example, Fischman, *The National Wildlife Refuges*.

51. On this latter point, see, for example, Sellars, *Preserving Nature in the National Parks*; Carr, *Wilderness by Design*.
52. US Fish and Wildlife Service, National Wildlife Refuge System, "About: Mission."
53. Interview with Michael Cramer, Literaturhaus Berlin, Berlin, Germany, September 19, 2013.

CHAPTER FOUR

1. The quote and following are taken from "History of Fisherman Island" interpretive sign text, located at Eastern Shore of Virginia National Wildlife Refuge, Virginia, author's visit, April 15, 2010.
2. US Fish and Wildlife Service, Fisherman Island, "Final Comprehensive Conservation Plan," 1–6.
3. Information on the refuge tours can be found at US Fish and Wildlife Service, Fisherman Island, "Refuge Tour." Hunters with tags for units on Fisherman Island are also allowed limited, seasonal access to parts of the refuge, and US Highway 13 effectively bisects the refuge.
4. Fischman, *The National Wildlife Refuges: Coordinating a Conservation System through Law*, 3.
5. The prospect of a distrustful or antagonistic public is far from hypothetical. Even as I write, a group of armed antigovernment protestors is occupying buildings at the Malheur National Wildlife Refuge in Oregon.
6. All numbers are in 1995 dollars. US Department of the Army, "Jefferson Proving Ground Final Environmental Impact Statement," 4-40 to 4-41.
7. Fischman, in *The National Wildlife Refuges*, 3, reports that 98 percent of all refuges allow some form of public access or recreation.
8. The first warden of the nation's first national wildlife refuge, Paul Kroegel at Pelican Island, reportedly patrolled for several years in the early twentieth century armed with his personal shotgun and a salary that ranged as *high* as $1 per month. See US Fish and Wildlife Service, National Wildlife Refuge System, *Fulfilling the Promise*.
9. 16 U.S.C. 715e.
10. US Fish and Wildlife Service, "Eastern Massachusetts National Wildlife Refuge Complex."
11. Interestingly, in this case even the FWS considers remediation as a liability that would cause more ecological harm than benefit.
12. The various refuge managers' anecdotal evidence of this is corroborated by lawsuits, agency memos, and the continued absence of remediation work at sites such as Rocky Mountain Arsenal, Big Oaks, Aroostook, and other M2W refuges.
13. US Fish and Wildlife Service, Shawangunk Grasslands, "History."
14. Rothbaum, "Wildlife Lives on Shawangunk Grasslands."
15. US Fish and Wildlife Service, Shawangunk Grasslands, "Visitor Opportunities."
16. US Fish and Wildlife Service, Shawangunk Grasslands, "Visitor Opportunities."
17. US Fish and Wildlife Service, Shawangunk Grasslands, "History."
18. US Fish and Wildlife Service, Shawangunk Grasslands, "History."
19. US Department of the Interior, "Budget Justifications and Performance Indication, Fiscal Year 2016: US Fish and Wildlife Service." The backlog was reduced substantially by short-term infusions from the 2009 federal stimulus (American Recovery and Reinvestment Act).

20. The Assabet River survey collected responses from a total of seventy-eight visitors. Some of these results are also presented in Havlick, "Restoration, History and Values at Transitioning Military Sites in the United States"; a related study appears in Havlick, Hourdequin, and John, "Examining Restoration Goals at a Former Military Site."
21. Other reasons included environmental education, to attend an organized event, to volunteer at the refuge, and to simply drop in out of curiosity while passing by.
22. The popularity of recreational uses such as bicycling and jogging present a dilemma for refuge managers. These activities are prohibited at some refuges due to concerns about disturbing wildlife. Assabet River managers share some of these concerns, but hope that people who come initially to recreate will come back to learn about the refuge's flora and fauna and appreciate its conservation mission more directly by spending time in the visitor center, participating in educational programs, or otherwise supporting refuge operations.
23. 16 U.S.C. §1.
24. In 2000, NPS director Robert Stanton issued an order intended to clarify conservation as the principal and overriding mission of the agency's mandate. By these terms, visitor services and public use remain a priority at national parks, but if these conflict with conservation then managers must recognize that the Park Service has "but a single purpose, namely, conservation." Stanton's order remains subject to the changes of subsequent administrations, however, and the George W. Bush administration actively worked to rewrite national park directives to privilege a wide array of public uses—regardless of their impact on conservation objectives; see Farquhar, "Revealed—Secret Changes to Park Rules." In June 2006, Secretary of the Interior Dirk Kempthorne signaled a return to prioritizing conservation when the dual mandates appear to conflict; for example, National Park Service, "Draft Management Policies to Guide the Park Service."
25. The 1997 act came in response to President Clinton's Executive Order 12996, which in turn responded to a series of Government Accountability Office reports that were critical of incompatible secondary uses on national wildlife refuges. Clinton's executive order described a mission for the National Wildlife Refuge System that was largely reiterated by Congress in the 1997 legislation.
26. National Wildlife Refuge System Improvement Act of 1997, Public Law 105-57, sec. 4, US Congress 111 Stat. 1252, October 9, 1997.
27. 65 Federal Register 33893 (2000). Following legislative acts, federal agencies publish implementing policies that provide a more detailed set of rules for how they will comply with the new legislation. For the Fish and Wildlife Service, these are developed in the Fish and Wildlife Service Manual, available online (and updated regularly as it responds to changing legislative and administrative requirements) at https://www.fws.gov/policy/manuals/. Fischman, in *The National Wildlife Refuges*, takes issue with this focus insofar as it discounts the importance of plant conservation despite the equal treatment of plants and animals in the 1997 legislation.
28. Fischman, *The National Wildlife Refuges*, 123; see also 163–82.
29. Fischman, *The National Wildlife Refuges*, 80.
30. See US Fish and Wildlife Service, Crab Orchard, "Comprehensive Conservation Plan"; see also the DVD *Around Crab Orchard* by Sarah Kanouse.
31. US Fish and Wildlife Service, Crab Orchard, "Draft Environmental Impact Statement/Comprehensive Conservation Plan," 327.

32. Mitchell, *The Lie of the Land*, 31.
33. US Department of the Army, "Disposal and Reuse of the Jefferson Proving Ground," 11.
34. US Department of the Army, "Disposal and Reuse of the Jefferson Proving Ground," 11.
35. Interview with Dan Matiatos, assistant refuge manager, Big Oaks National Wildlife Refuge, December 14, 2005.
36. Interview with Matiatos, December 14, 2005.
37. US Department of the Army, "Disposal and Reuse of the Jefferson Proving Ground," 11.
38. "Memorandum of Agreement," 2.
39. *Atlanta Journal and Constitution*, "Pilot Safely Ejects before Jet Crashes."
40. Similar conflicts over military activities as "secondary uses" of national wildlife refuges emerged from the GAO reports discussed in more depth in chapter 2; see also US Government Accounting Office, "National Wildlife Refuges: Continuing Problems."
41. Interview with an anonymous FWS official, December 16, 2005.
42. See minutes from the Big Oaks National Wildlife Refuge Restoration Advisory Board, April 30, 2003, 75–100. There is no indication that the army has plans to add to the quantity of DU currently present on-site at Big Oaks. In 2013, the army submitted a proposal for the DU firing range to the US Nuclear Regulatory Commission decommission, which has the authority to release the army from management and liability. The army withdrew this request in 2016. See the site report: US Nuclear Regulatory Commission, "Jefferson Proving Ground."
43. See, for example, Ulrich Beck, *Ecological Politics in an Age of Risk*, *Risk Society: Towards a New Modernity*, and *World Risk Society*.
44. Beck, *World Risk Society*, 2. In addition to the two stages of modernity, Beck describes a third stage of "pre-industrial society" typical of traditional societies. See also Beck, *Ecological Politics in an Age of Risk* and *Risk Society*; and Mythen, *Ulrich Beck*.
45. Interview with Ed Anderson, biologist, Savanna Field Station, Illinois Department of Natural Resources, May 30, 2006.
46. Admittedly, the entry process at Big Oaks likely instills a certain cautiousness in most visitors.
47. Beck, "Risk Society and the Provident State," 27.
48. Beck, "Risk Society and the Provident State," 37.
49. These same risks of transportation were deemed acceptable when it came to shipping radioactive waste to the Department of Energy's Yucca Mountain site in Nevada.
50. *Engineering News Record*, "Asphaltic Membrane Is Used"; *Omaha* (Nebraska) *World-Herald*, "Leakproof Bottom Underlies Lake."
51. Beck, *World Risk Society*, 88.
52. Beck, in *World Risk Society*, 5, notes a similar opening of decision-making that comes with the onset of risk society and suggests that institutional reform of private corporations and the sciences (and I would add, federal agencies) "could encourage environmental innovations and help to construct a better developed public sphere in which crucial questions of value that underpin risk conflicts can be debated and judged."
53. Eisenhower quoted in Barnett, *The Meaning of Environmental Security*, 96.
54. See Lutz, *The Bases of Empire*; and de Yoanna, "Pattern of Misconduct." De Yoanna's article describes heightened levels of suicide, substance abuse, and drunk driving

among soldiers returning from active duty in Iraq, many of whom have been diagnosed as suffering from post-traumatic stress disorder but instead of receiving institutional support from the army face dishonorable discharges and a loss of employment and health and education benefits. For a more detailed consideration of post-traumatic stress, see Hautzinger and Scandlyn, *Beyond Post-Traumatic Stress*.

55. Beck refers to this as "normal birth" (*World Risk Society*, 50).
56. Beck, *World Risk Society*, 61.
57. See, for example, Sasha Davis, "Fish and Wildlife Is Another Name for the Navy" and "Representing Place."
58. See, for example, Woodward, *Military Geographies*, 54.

CHAPTER FIVE

1. Sola, "Germany to Turn 62 Military Bases into Nature Sanctuaries."
2. Michael Cramer, "Iron Curtain Trail"; Iron Curtain Trail, "The Iron Curtain Trail: Experiencing the History of Europe's Division"; see also Hammer, "A Barrier Gone, but Not Forgotten."
3. Cramer, *Iron Curtain Trail, Part 3, Iron Curtain Trail, Part 2*.
4. Interview with Michael Cramer, Literaturhaus Berlin, Berlin, Germany, September 19, 2013.
5. Cramer, "Iron Curtain Trail."
6. Cramer, *Iron Curtain Trail, Part 2*, 240.
7. See Bičík and Štěpánek, "Post-War Changes in Bohemia and Moravia"; and Bičík, Jeleček, and Štěpánek, "Land-Use Changes in Czechia."
8. Bičík, Jeleček, and Štěpánek, "Land-Use Changes in Czechia."
9. Šumava National Park. "Nature: Basic Information."
10. Bičík and Štěpánek, "Post-War Changes in Bohemia and Moravia."
11. Kupková et al., "Land Cover Changes along the Iron Curtain."
12. Higuchi et al., "Satellite Tracking of White-Naped Crane Migration"; Ke Chung Kim, "Preserving Biodiversity in Korea's Demilitarized Zone"; Associated Press, "Turner: Turn DMZ into a Peace Park"; Thomas, "The Exquisite Corpses of Nature and History." See also Weisman, *The World without Us*; and Card, "Paradise in No Man's Land."
13. Card, "Paradise in No Man's Land."
14. Card, "Paradise in No Man's Land"; Westing, "Toward Environmental Sustainability on the Korean Peninsula"; Platt, "South Korea Seeks to Protect Species."
15. Shin, "Peace Park Tourism Planning and Development."
16. Associated Press, "Ted Turner: Turn DMZ into a Peace Park."
17. Azios, "Demilitarized Zone Now a Haven"; Jungyun, "An Oasis of Wildlife"; Wall, "Korean DMZ Teems with Wildlife."
18. Cain, "South Korea Tries to Re-brand DMZ."
19. Korea Tourism Organization, "Demilitarized Zone Tours."
20. Korea Tourism Organization, "Demilitarized Zone Tours."
21. DMZ Tours, http://www.dmztours.com/.
22. Korea Tourism Organization, "Demilitarized Zone Tours."
23. Sullivan, "Borderline Absurdity DMZ." See also Shin, "A Critical Review of Peace Park Tourism Planning," 115.
24. Quoted in Lee, "A New Paradigm for Trust-Building on the Korean Peninsula."
25. Pearson and Ingrassia, "Tank Traps Meet Tourist Traps."
26. Korea.net News; Gyeonggi G-News, "Gyeonggi Province Adds Stories to Pyeonghwa Nuri-gil."

27. Gyeonggi G-News, "Gyeonggi Province Adds Stories to Pyeonghwa Nuri-gil."
28. Korea.net News.
29. Korea DMZ Peace-Life Valley, "Greetings."
30. See Brady, "Life in the DMZ"; Westing, "Toward Environmental Sustainability on the Korean Peninsula"; Shin, "A Critical Review of Peace Park Tourism Planning."
31. Westing, "The DMZ as a Bridge between the Two Koreas."
32. Brady, "Life in the DMZ," 610.
33. Government of Western Australia, Parks and Wildland Service, "Montebello Islands Marine Park."
34. See Leonard, "Tampering with History."
35. McClelland, *The Report into British Nuclear Tests in Australia*, 109.
36. In addition to maintaining a website about the Montebello Islands marine park, Western Australia also publishes a park newsletter (see, for example, http://parks.dpaw.wa.gov.au/sites/default/files/downloads/parks/Montebellos%20News%20Issue%202%20July%202015.pdf) and an information guide.
37. Government of Western Australia, Parks and Wildland Service, "Montebello Islands Marine Park."
38. McClelland, *The Report into British Nuclear Tests in Australia*, 261.
39. McClelland, *The Report into British Nuclear Tests in Australia*, 118.
40. Preparatory Commission for the Comprehensive Nuclear-Test-Ban Treaty Organization, "The United Kingdom's Nuclear Testing Programme."
41. Preparatory Commission for the Comprehensive Nuclear-Test-Ban Treaty Organization, "The United Kingdom's Nuclear Testing Programme."
42. Preparatory Commission for the Comprehensive Nuclear-Test-Ban Treaty Organization, "The United Kingdom's Nuclear Testing Programme"; Kent Acott, "Dark Cloud Hangs over Atomic Test."
43. McClelland, *The Report into British Nuclear Tests in Australia*, 121.
44. Grabosky, "A Toxic Legacy."
45. See Sasha Davis, *The Empires' Edge*, 66; Smith and Moore, "Radiological Clean-Up of Bikini Atoll."
46. Smith and Moore, "Radiological Clean-Up of Bikini Atoll," 3.
47. The multiple dislocations and conditions faced by Bikinians-in-exile remains an ongoing and tragic story. See, for example, Niedenthal, "A History of the People of Bikini Following Nuclear Weapons Testing."
48. See Marshall Islands Dose Assessment & Radioecology Program, "Bikini Atoll."
49. Sasha Davis, *The Empires' Edge*, 68.
50. Marshall Islands Dose Assessment & Radioecology Program, "Bikini Atoll."
51. UNESCO, "Bikini Atoll Nuclear Test Site."
52. Abell, "Bikini Is Da Bomb."
53. The Bikiniatoll.com website also includes information about the dislocation of the native Bikini islanders; Lust4Rust, "Bikini Atoll," posts similar claims for Bikini as the ultimate in wreck diving.
54. For more-extended treatments of some of these concerns, see Hourdequin and Havlick, *Restoring Layered Landscapes*.
55. Regarding tourist motivations to visit the DMZ, see, for example, Bigley et al., "Motivations for War-Related Tourism"; on tourism along the Iron Curtain, see, for example, Eckert, "Greetings from the Zonal Border."
56. Eckert, "Greetings from the Zonal Border."
57. Indies Trader, "Diving Crystal Clear Water."

58. Indies Trader, Diving Crystal Clear Water."
59. Pearson and Ingrassia, "Tank Traps Meet Tourist Traps."
60. European Green Belt, "Memorial Landscape."
61. Smale, "Migrants Race North as Hungary Builds a Fence."
62. See, for example, Löwenheim, *The Politics of the Trail.*
63. See, for example, Michael Cramer, *Iron Curtain Trail, Part 1, Iron Curtain Trail, Part 2, Iron Curtain Trail, Part 3;* Hammer, "A Barrier Gone, but Not Forgotten"; Schwägerl, "Green Belt Rises in Germany"; Lamarra, "Riding the Iron Curtain."
64. For lucid arguments on how ruination can open up productive new forms of meaning and interpretation, see DeSilvey and Edensor, "Reckoning with Ruins"; and Edensor, *Industrial Ruins.* For a consideration of more carefully designed memorials and the potential these may have for constraining meaning, see Gobster, "The 'Museumification' of Nature."
65. Kwon, *One Place after Another,* 6.
66. Verheyen, *United City, Divided Memories?,* 202.

CHAPTER SIX

1. On April 26, 2017, President Donald Trump signed an executive order calling for an evaluation of all recent national monument designations larger than 100,000 acres. It remains to be seen, as of July 2017, whether this order will result in a reduction in size of any of these monument lands, including the Pacific Remote Islands. See Eilperin, "Trump Orders Review of National Monuments."
2. US White House, "Obama to Designate Largest Marine Monument."
3. US Fish and Wildlife Service, "Obama Protects Vital Marine Habitat."
4. See Vine and Pemberton, "Marine Protection as Empire Expansion." For a related critique, see Harris, "Militarism in Environmental Disguise."
5. Maclellan, "Cleaning Up Johnston Atoll."
6. The term *military environmentalism* comes from Coates et al., "Defending Nation, Defending Nature?"; *khaki conservation* originates from Woodward, "Khaki Conservation"; and *ecological militarization* is used in Havlick, "Disarming Nature," and Havlick, "Logics of Change for Military-to-Wildlife Conversions."
7. See, for example, Shields, "Defence of Nature's Realm."
8. Ward, "Bombing Range Is National Example."
9. Center for Environmental Management of Military Lands (CEMML), accessed June 26, 2016, https://www.cemml.colostate.edu/cultural/09476/cemmliraqenl.html.
10. CEMML.
11. See, for example, Weeks, "From Bombs to Birds."
12. This is also the name of the Rocky Flats Cold War Museum's (http://www.rockyflatsmuseum.org/) newsletter.
13. Meierotto, "A Disciplined Space," 640.
14. For a specific example of this blending of conservation and military objectives, see Meierotto, "A Disciplined Space."
15. The US Air Force "Energy, Environment, Safety, and Occupational Health" sustainability report, for example, described the air force's "one common foundation—persistent, lethal, overwhelming air, space, and cyberspace power massed and able to be brought to bear anywhere at any time" (1).
16. See US Department of Defense, "Defense Environmental Programs Annual Report to Congress for FY 2014." The 2016 FY request is the lowest since 2010, when the environmental programs budget was $4.5 billion.

17. See Sierra Club Foundation, Annual report, 2014. The Nature Conservancy—the US conservation group with the largest assets—reported approximately $565 million in conservation-related expenses in 2015, and total support and revenues of $948 million; see Nature Conservancy, "Our World: 2015 Annual Report."
18. See US Department of Defense, "Environmental Programs Annual Report FY 2014," table 1 (p. 2); note that FY 2014 data were included in the publication dated September 2015.
19. US Department of Defense, "Environmental Programs Annual Report FY 2014," 16.
20. US Department of Defense, "Environmental Programs Annual Report FY 2014," 10.
21. US Department of the Army, "Army Strategy for the Environment."
22. US Air Force, "Energy, Environment, Safety and Occupational Health."
23. Secretary Mabus address to the Naval Energy Forum, October 14, 2009, McLean, Virginia, as cited in Pew Project on National Security, Energy, and Climate, "Reenergizing America's Defense."
24. US Department of the Army, "Army Earth Day Message 2010." For Net Zero initiatives, see, for example, US Department of the Army, "Sustainability Report 2014."
25. US Department of the Army, "Earth Day 2016: Army Message."
26. US Air Force, "Energy Strategic Plan to a Net Zero Installation."
27. Interview with James B. Pocock, professor, Civil and Environmental Engineering, US Air Force Academy, Colorado Springs, CO, November 9, 2009; US Air Force, "Energy Strategic Plan to a Net Zero Installation."
28. Baillie, "Academy Hits Conservation Target."
29. US Department of the Army, Fort Carson Public Affairs Office, "Fort Carson Statistics"; interview with Adam Ozier, assistant sustainability coordinator, Fort Carson, CO, March 2010.
30. Interview with Ozier; Roper, "Fort Carson Lands Copter Brigade"; 50 percent increase in building square footage reported in National Renewable Energy Laboratory, "Helping Fort Carson Meet DOD Energy Goals."
31. Colorado Renewable Energy Society. "Sustainable Fort Carson." See also US Department of the Army, "Fort Carson 2014 Sustainability Report."
32. Interview with Mary J. Barber, Fort Carson Installation Sustainability Resource Officer, Fort Carson, Colorado, October 7, 2015.
33. US Department of the Army, "Fort Carson 2014 Sustainability Report."
34. Juniper, "Fort Carson's Extraordinary Sustainability Journey"; Havlick, "Militarization, Conservation, and U.S. Base Transformations"; Havlick and Perramond, "Militarized Spaces and Open Range."
35. Juniper, "Fort Carson, the Mountain Post SEMS." See also "Fort Carson 2014 Sustainability Report."
36. "Fort Carson 2014 Sustainability Report."
37. US Government Accountability Office, "DOD Efforts Regarding Net Zero Goals," 11.
38. US Government Accountability Office, "DOD Efforts Regarding Net Zero Goals," 11.
39. US Government Accountability Office, "DOD Efforts Regarding Net Zero Goals," 12.
40. See, for example, Hautzinger and Scandlyn, *Beyond Post-Traumatic Stress*; Sanders, *Green Zone*; Lutz, *The Bases of Empire*; Philipps, *Lethal Warriors*.
41. See Hajer, *The Politics of Environmental Discourse*, 25–26. More generally on ecological modernization, see Mol, "Ecological Modernisation and Institutional Reflexivity" and "The Environmental Movement in an Era of Ecological Modernisation"; Mol and Sonnenfeld, "Ecological Modernisation around the World"; Mol and Spaargaren,

"Ecological Modernisation Theory in Debate"; and Gonzalez, "Local Growth Coalitions and Air Pollution Controls."

42. Christoff, "Ecological Modernisation, Ecological Modernities," 497. See also Hajer, "Ecological Modernisation as Cultural Politics"; Harvey, *Justice, Nature and the Geography of Difference*; Dryzek, *The Politics of the Earth*; and Gonzalez, "Local Growth Coalitions and Air Pollution Controls," which each raise versions of these critiques.
43. Dryzek, *The Politics of the Earth*, 179.
44. See, for example, Ke Chung Kim, "Preserving Biodiversity in Korea's Demilitarized Zone"; Leslie, Meffe, and Hardesty, *Conserving Biodiversity on Military Lands*; Hoffecker, *Twenty-Seven Square Miles*; Schmidt, "Nature Sows Life Where Man Brewed Death"; Hoosier Environmental Council, "Proposal for Jefferson Proving Ground in Trouble!"; Rocky Mountain Arsenal National Wildlife Refuge Act of 1991.
45. This echoes former presidential candidate Dennis Kucinich's proposal to create a Department of Peace. A "wishful" federal agency proposed/created by artist Sarah Kanouse and geographer Shiloh Krupar also offers one stimulating alternative to this business-as-usual scenario. Information about the National Toxic Land/Labor Conservation Service (National TLC Service) is available at http://www.nationaltlcservice.us/.
46. Lapham, "Lionhearts."
47. Office of Budget and Management Department of Defense budget, FY 2007.
48. See, for example, Cheney, "Defense Strategy for the 1990s."
49. Vine, *Base Nation*; Sanders, *Green Zone*.
50. Ackerman, "Afghan 'Burn Pit' Could Wreck Troops' Hearts, Lungs." See also Vine, *Base Nation*, 138.
51. Langhelle, "Why Ecological Modernization and Sustainable Development Should Not Be Conflated," 309. See also Mol and Spaargaren, "Environment, Modernity, and the Risk-Society"; Dryzek, *The Politics of the Earth*; and Gonzalez, "Local Growth Coalitions and Air Pollution Controls."
52. US Fish and Wildlife Service, Big Oaks, "Fire Management Plan: Appendix K," 25.
53. See, for example, Philipps, "Veterans Feel Cost of U.S. Nuclear Tests."
54. Irwin and Wynne, introduction to *Misunderstanding Science?*
55. Irwin and Wynne, introduction to *Misunderstanding Science?*
56. Leopold, *A Sand County Almanac*, 223.
57. Leopold, *A Sand County Almanac*, 223.
58. See Shulman, *The Threat at Home*; and Sanders, *Green Zone*.
59. US Environmental Protection Agency, "The Environmental Challenge of Military Munitions and Federal Facilities."
60. Reuben, "Reducing Environmental Cancer Risk."
61. Interview with Roger Shakely, Rocky Mountain Arsenal project manager, Shell Oil Company, July 21, 2004, Commerce City, Colorado. The same is true of the disposal of nuclear waste: The planning horizon for the proposed federal Yucca Mountain nuclear depository is ten thousand years. See Kuletz, *The Tainted Desert*; and US Department of Energy, Office of Civilian Radioactive Waste Management, "Final Environmental Impact Statement for a Geologic Repository at Yucca Mountain."
62. See Blowers, "Environmental Risk Policy."
63. Cohen makes a related point in "Risk Society and Ecological Modernisation," 106.
64. Christoff, "Ecological Modernisation."
65. Christoff, "Ecological Modernisation," 496.
66. See Cohen, "Risk Society and Ecological Modernisation," especially 108.

67. Details on the Loring Air Force Base and subsequent management as Aroostook National Wildlife Refuge come from interviews conducted during a site visit in July 2008 and follow-up conversations in June 2012 with refuge officials.
68. US Fish and Wildlife Service, Aroostook, "Auto Tour Route."
69. For coverage on the igloos-as-hibernacula story, see US Fish and Wildlife Service, "Retired Bunkers Used as Bat Hibernacula at Aroostook"; and Platt, "To the Bat Bunker!"
70. See US Fish and Wildlife Service, Aroostook, "Plan Your Visit."
71. As above, details on the Loring Air Force Base and subsequent management as Aroostook National Wildlife Refuge come from interviews conducted during a site visit in July 2008, and follow-up conversations in June 2012 with refuge officials.
72. The site in question was the Lost Mound Unit of the Upper Mississippi River National Wildlife and Fish Refuge, or formerly, the Savanna Army Depot in northwestern Illinois.
73. Louden, "Army Expansion Plans Have Ranchers on Edge"; de Yoanna, "Targeting Paradise."
74. Louden, "Plans Have Ranchers on Edge," 1E; see also Havlick and Perramond, "Militarized Spaces and Open Range."
75. De Yoanna, "Targeting Paradise"; Louden, "Plans Have Ranchers on Edge."
76. "Piñon Canyon Expansion Opposition Coalition, "Our Heritage, Our History."
77. US Department of Defense. "2015 Secretary of Defense Environmental Awards" (includes a list of previous winners). The National Wildlife Federation's award is cited at GlobalSecurity.org. "Piñon Canyon Maneuver Site (PCMS)."
78. Kuletz, *The Tainted Desert*, 77.
79. US Department of Defense, "Base Realignment and Closures," 20.

CHAPTER SEVEN

1. Altman, "Best Cities for Veterans to Live."
2. Meyerson, *Nature's Army*.
3. Jacoby, *Crimes against Nature*.
4. Nguyen, *Nothing Ever Dies*, 2–3.
5. Many of these same questions and problems of course pertain to many parts of Germany, Vietnam and Southeast Asia, Iraq, Afghanistan, Syria, and other places that have suffered broad impacts of modern warfare. My focus on Japan is intended as a case study to highlight an array of issues that, despite certain unique characteristics, surely apply to many other heavily impacted settings.
6. See, for example, PBS *Frontline*'s "Holocaust Education in Germany: An Interview," May 19, 2005, accessed January 30, 2017, http://www.pbs.org/wgbh/pages/frontline/shows/germans/germans/education.html.
7. The same is true of Nagasaki, site of the only other atomic bomb used in war, though seemingly to a lesser extent. See, for example, Rich, "What about Us, Nagasaki Asks."
8. See, for example, Yoneyama, *Hiroshima Traces*, and "Memories in Ruins."
9. Yoneyama, "Memories in Ruins," 188.
10. Sasha Davis makes a similar point in "Military Natures."
11. Rocky Mountain Arsenal National Wildlife Refuge Act of 1991.
12. Cronon, "The Trouble with Wilderness," 69–90. For further consideration of Cronon's concerns, see Baird and Nelson, *The Great New Wilderness Debate*; and Havlick, "Reconsidering Wilderness."
13. Cronon, "The Trouble with Wilderness," 80.
14. Misrach, *Bravo 20*.

15. Misrach, *Bravo 20*, 95.
16. Misrach, *Bravo 20*, 95.
17. Interview (by phone) with William Kolodnicki, then Aroostook refuge manager, July 6, 2006.
18. See, for example, the Forest History Society's motto, "By understanding our past, we shape our future" (https://foresthistory.org/).
19. Foote, *Shadowed Ground*, 357.
20. Beck, in *World Risk Society*, 50, calls this the "calculus of risks" and maintains that the DOD and other institutions have, in fact, produced uninsurable hazards.
21. Historian Jacob Hamblin presents a related though conversely framed perspective about the weaponization of nature in his thought-provoking book, *Arming Mother Nature: The Birth of Catastrophic Environmentalism*.
22. National Trust, "Orford Ness National Nature Reserve."
23. Much of the technical history of the site is drawn from Cocroft and Alexander, "Atomic Weapons Research Establishment, Orford Ness, Suffolk."
24. National Trust, "Orford Ness."
25. Sophia Davis, "Military Landscapes and Secret Science."
26. See Sophia Davis, "Military Landscapes and Secret Science," 147.
27. As DeSilvey and Edensor point out in "Reckoning with Ruins," 473, the decision to leave structures in place to fall to ruin is also influenced by a "state of limbo" where military infrastructure is "not allowed to be demolished but not considered valuable enough to merit expenditure on stabilization."
28. US Fish and Wildlife Service, Caddo Lake, "Plan Your Visit."
29. Available at the National Toxic Land/Labor Conservation Service website, http://www.nationaltlcservice.us/.
30. National TLC Service brochure.
31. The quote "plowed up, shredded, and buried" is from "Uravan, Colorado: The Town That Was," a brochure produced by the Rimrocker Historical Society, Nucla, Colorado; "lives, lands, and bodies" is from "National Cold War Monuments and Environmental Heritage Trail: A Design Charrette" booklet, provided for the March 19, 2016, workshop organized by the National TLC Service.
32. For much more of the Rocky Flats story, see Iverson, *Full Body Burden*.
33. Candelas, "Candelas: Life Wide Open."
34. Candelas, "Rocky Flats National Wildlife Refuge."
35. National Park Service, Manhattan Project, "Learn about the Park."
36. US Fish and Wildlife Service, Hanford Reach, "Habitats."
37. US White House, "Establishment of the Hanford Reach National Monument."
38. Cram, "Wild and Scenic Wasteland."
39. Murray, "Senator Murray Applauds Hanford Reach Designation."
40. See Kelly, "The Making of the Manhattan Project Park."
41. See Kelly, "The Making of the Manhattan Project Park"; and Atomic Heritage Foundation, "Preserving the Manhattan Project Sites for Future Generations."
42. Nguyen, *Nothing Ever Dies*, 10–11.

BIBLIOGRAPHY

Abell, John C. "May 21, 1956: Bikini Is Da Bomb." *Wired*, May 20, 2010. http://www.wired.com/2010/05/0521bikini-h-bomb/.

Ackerman, Spencer. "Leaked Memo: Afghan 'Burn Pit' Could Wreck Troops' Hearts, Lungs." *Wired*, May 22, 2017. https://www.wired.com/2012/05/bagram-health-risk.

Acott, Kent. "Dark Cloud Hangs over Atomic Test." *West Australian*, October 3, 2012. https://au.news.yahoo.com/thewest/wa/a/15019060/dark-cloud-hangs-over-atomic-test/.

Alameda Point Environmental Report. "Film: 'Demilitarized Landscapes' Produced for Oakland Museum." October 15, 2014. https://alamedapointenvironmentalreport.wordpress.com/2014/10/15/film-demilitarized-landscapes-produced-for-oakland-museum/.

Alameda Point Info. "Frequently Asked Questions: What Is Happening with the Wildlife Refuge?" Accessed November 25, 2014. http://alamedapointinfo.com/faq/what-happening-wildlife-refuge.

Alamogordo Daily News. "Trinity Site to Be Open April 4." *Albuquerque Journal*, March 8, 2015. http://www.abqjournal.com/551635/news/trinity-site-to-be-open-april-4.html.

Allard, Wayne. Introduction of legislation establishing Colorado Metropolitan Wildlife Refuge. US House of Representatives, H5436-H5437, July 15, 1991. Available at https://www.gpo.gov/fdsys/pkg/GPO-CRECB-1991-pt13/pdf/GPO-CRECB-1991-pt13-2.pdf, page 2.

Altman, George. "What Are the Best Cities for Veterans to Live?" *USA Today*, June 16, 2015. http://www.usatoday.com/story/news/nation-now/2015/06/16/cities-military-veterans-virginia-beach-alexandria-bethesda/28808117/.

Anderson, Reginald, and David G. Havlick. "History and Values in Ecological Restoration Workshop." *Ecological Restoration* 31 (2013): 7–10.

Armstrong, J., J. R. McDermott, and J. D. Ripley. "The U.S. Air Force Embraces Ecosystem Management through a Landscape Assessment in the Front Range of the Rocky Mountains." *Federal Facilities Environmental Journal* 11, no. 3 (Autumn 2000): 125–34.

Associated Press. "Crews Find No New Signs of Chemical Weapon at Arsenal." *Denver Post*, November 19, 2007. http://www.denverpost.com/2007/11/19/crews-find-no-new-signs-of-chemical-weapon-at-arsenal/.

———. "Ted Turner: Turn DMZ into a Peace Park." *USA Today*, November 18, 2005. http://usatoday30.usatoday.com/news/world/2005-11-18-turnerdmz_x.htm.

Atlanta Journal and Constitution. "Pilot Safely Ejects Before Jet Crashes." November 18, 1998, 7A.

Atomic Heritage Foundation. "Preserving the Manhattan Project Sites for Future Generations." Accessed June 21, 2016. http://www.atomicheritage.org/preserving-manhattan-project-sites-future-generations.

Azios, Tony. 2008. "Korean Demilitarized Zone Now a Wildlife Haven." *Christian Science Monitor*, November 21, 2008. http://www.csmonitor.com/Environment/Wildlife/2008/1121/korean-demilitarized-zone-now-a-wildlife-haven.

Baillie, Amber. "Academy Hits Conservation Target." US Air Force Academy Public Affairs, September 30, 2014. http://www.af.mil/News/ArticleDisplay/tabid/223/Article/502988/academy-hits-conservation-target.aspx.

Barnett, Jon. *The Meaning of Environmental Security: Ecological Politics and Policy in the New Security Era*. London: Zed Books, 2001.

Bearden, David. "Vieques and Culebra Islands: An Analysis of Cleanup Status and Costs." Congressional Research Service Report for Congress, July 2005.

Beck, Ulrich. *Ecological Politics in an Age of Risk*. Cambridge, UK: Polity Press, 1995.

———. "Risk Society and the Provident State." In *Risk, Environment and Modernity: Towards a New Ecology*, edited by Scott Lash, Bronislaw Szerszynski, and Brian Wynne, 27–43. London: Sage Publications, 1996.

———. *Risk Society: Towards a New Modernity*. Translated from German by Mark Ritter. London: Sage Publications, 1992.

———. *World Risk Society*. Cambridge, UK: Polity Press, 1999.

Becker, Marc. "Vieques: Long March to People's Victory." *Against the Current*, November/December 2003. https://www.solidarity-us.org/node/503.

Benton, Nancy, J. Douglas Ripley, and Fred Powledge, eds. *Conserving Biodiversity on Military Lands: A Guide for Natural Resource Managers*. 2008 edition. Arlington, VA: Nature Serve.

Berry, Wendell. "The Futility of Global Thinking." In *Learning to Listen to the Land*, edited by Bill Willers, 150–56. Washington, DC: Island Press, 1991.

Bičík, Ivan, Leoš Jeleček, and Vít Štěpánek. "Land-Use Changes and Their Social Driving Forces in Czechia in the 19th and 20th Centuries." *Land Use Policy* 18, no. 1 (2001): 65–73.

Bičík, Ivan, and Vít Štěpánek. "Post-War Changes of the Land-Use Structure in Bohemia and Moravia: Case Study Sudetenland," *GeoJournal* 32, no. 3 (1994): 253–59.

Bigley, James D., Choong-Ki Lee, Jinhyung Chon, and Yooshik Yoon. "Motivations for War-Related Tourism: A Case of DMZ Visitors in Korea." *Tourism Geographies* 12, no. 3 (2010): 371–94.

Blowers, Andrew. "Environmental Risk Policy: Ecological Modernisation or the Risk Society?" *Urban Studies* 34, no. 5–6 (1997): 845–71.

Botkin, Daniel B. *Discordant Harmonies: A New Ecology for the Twenty-First Century*. New York: Oxford University Press, 1990.

Bradshaw, Anthony. "What Do We Mean by Restoration?" In *Restoration Ecology and Sustainable Development*, edited by Krystyna M. Urbanska, Nigel R. Webb, and Peter J. Edwards, 8–13. Zurich: Swiss Federal Institute of Technology, 1997.

Brady, Lisa M. "Life in the DMZ: Turning a Diplomatic Failure into an Environmental Success." *Diplomatic History* 32, no. 4 (2008): 585–611.

Broad, William J. "Bid to Preserve Manhattan Project Sites in a Park Stirs Debate." *New York Times*, December 4, 2012, D3.

Burger, Joanna. "Integrating Environmental Restoration and Ecological Restoration: Long-Term Stewardship at the Department of Energy." *Environmental Management* 26 (2000): 469–578.

Burger, Joanna, T. M. Leschine, M. Greenberg, J. R. Karr, M. Gochfeld, and C. W. Powers. "Shifting Priorities at the Department of Energy's Bomb Factories: Protecting Human and Ecological Health." *Environmental Management* 31 (2003): 157–67.

Caddo Lake Institute. "Our Mission." Last updated 2008. http://www.caddolakeinstitute.us/index.html.

Cain, Geoffrey. "South Korea Tries to Re-brand DMZ as Rare Animal Sanctuary." NBC News, October 30, 2014. http://www.nbcnews.com/news/asian-america/south-korea-tries-re-brand-dmz-rare-animal-sanctuary-n231641.

Callicott, J. Baird, and Michael P. Nelson, eds. *The Great New Wilderness Debate*. Athens, GA: University of Georgia Press, 1998.

Candelas. "Candelas: Life Wide Open." Accessed June 18, 2016. http://www.candelaslife.com/.

———. "Rocky Flats National Wildlife Refuge." Accessed June 21, 2016. http://www.candelasrockyflats.com/.

Card, James. "Paradise in No Man's Land." *Earth Island Journal*, Spring 2008. http://www.earthisland.org/journal/index.php/eij/article/paradise_in_no_mans_land/.

Carr, Ethan. *Wilderness by Design: Landscape Architecture and the National Park Service*. Lincoln: University of Nebraska Press, 1999.

Castree, Noel. "Commentary." *Environment and Planning A* 36 (2004): 191–94.

———. *Nature*. New York: Routledge, 2005.

Cheney, Dick. "Defense Strategy for the 1990s: The Regional Defense Strategy." US Department of Defense, 1993.

Choi, Young D. "Restoration Ecology to the Future: A Call for New Paradigm." *Restoration Ecology* 15, no. 2 (2007): 351–53.

———. "Theories for Ecological Restoration in Changing Environment: Toward 'Futuristic' Restoration." *Ecological Research* 19, no. 1 (2004): 75–81.

Christoff, Peter. "Ecological Modernisation, Ecological Modernities." *Environmental Politics* 5 (1996): 476–500.

Coates, Peter. "Borderland, No-Man's Land, Nature's Wonderland: Troubled Humanity and Untroubled Earth." *Environment and History* 20 (2014): 499–516.

Coates, Peter, Tim Cole, Marianna Dudley, and Chris Pearson. "Defending Nation, Defending Nature? Militarized Landscapes and Military Environmentalism in Britain, France, and the United States." *Environmental History* 16, no. 3 (2011): 456–91.

Cocroft, Wayne, and Magnus Alexander. "Atomic Weapons Research Establishment, Orford Ness, Suffolk: Cold War Research and Development Site." Research Department Report Series 10-2009. English Heritage, Portsmouth: 2009.

Cohen, Maurie J. "Risk Society and Ecological Modernisation," *Futures* 29, no. 2 (1997): 105–19.

Colón-Ramos, Daniel. "Letter to President Barack Obama about Vieques." *Huffington Post*, April 20, 2015. http://www.huffingtonpost.com/daniel-coloramos/letter-to-president-barack-obama-about-vieques_b_7100090.html.

Colorado Renewable Energy Society. "Sustainable Fort Carson." September 27, 2014. http://secres.org/sustainable-fort-carson.

Cowan, Tadlock. "Military Base Closures: Socioeconomic Impacts." Congressional Research Service, RS22147. February 7, 2012.

Cram, Shannon. "Wild and Scenic Wasteland: Conservation Politics in the Nuclear Wilderness." *Environmental Humanities* 7, no. 1 (2015): 89–105.

Cramer, Michael. "Iron Curtain Trail." The Greens/EFA in the European Parliament. Accessed December 3, 2013. http://www2.ironcurtaintrail.eu/uploads/brochure_iron_curtain_trail.pdf. Brochure.

———. *Iron Curtain Trail, Part 1*. Rodingersdorf, Austria: Verlag Esterbauer, 2007.
———. *Iron Curtain Trail, Part 2*. Rodingersdorf, Austria: Verlag Esterbauer, 2012.
———. *Iron Curtain Trail, Part 3*. Rodingersdorf, Austria: Verlag Esterbauer, 2010.

Cronon, William. "The Trouble with Wilderness; or, Getting Back to the Wrong Nature." In *Uncommon Ground: Toward Reinventing Nature*, edited by W. Cronon, 69–90. New York: W. W. Norton, 1995.

Davis, Sasha. *The Empires' Edge: Militarization, Resistance, and Transcending Hegemony in the Pacific*. Athens, GA: University of Georgia Press, 2015.

———. "'Fish and Wildlife Is Another Name for the Navy': Military Destruction, Environmental Preservation and Social Justice." Presentation at the Annual Meeting of the Association of American Geographers, Chicago, Illinois, March 2006.

———. "Military Natures: Militarism and the Environment." *GeoJournal* 69, no. 3 (2007): 131–34.

———. "Representing Place: 'Deserted Isles' and the Reproduction of Bikini Atoll." *Annals of the Association of American Geographers* 95, no. 3 (2005): 607–25.

Davis, Sophia. "Military Landscapes and Secret Science: The Case of Orford Ness." *Cultural Geographies* 15 (2008): 143–49.

DENIX (DOD Environment, Safety and Occupational Health Network and Information Exchange). "Fiscal Year 2009 Secretary of Defense Environmental Award: Vieques Naval Installation: Environmental Restoration—Individual/Team." Accessed July 3, 2017. http://www.denix.osd.mil/awards/fy09secdef/erit/project-management-team-vieques-naval-installation-puerto-rico/.

———. "The 2012 Secretary of Defense Environmental Awards 50th Anniversary." http://www.denix.osd.mil/awards/upload/2012_Environ_Awards_Brochure_online.pdf (webpage discontinued).

———. "2015 Secretary of Defense Environmental Awards: About the Awards." http://www.denix.osd.mil/awards/fy14secdef/awards-information/about-the-2015-secretary-of-defense-environmental-awards/.

DeSilvey, Caitlin, and Tim Edensor. "Reckoning with Ruins." *Progress in Human Geography* 37, no. 4 (2012): 465–85.

de Yoanna, Michael. "Army Maneuvers." *Colorado Springs Independent*, August 24–30, 2006, 11.

———. "Pattern of Misconduct: Fort Carson Soldiers Allege Abuse and Intimidation." *Colorado Springs Independent*, July 13–19, 2006, 14–19.

———. "Targeting Paradise." *Colorado Springs Independent*, August 10–16, 2006, 14–17.

DMZ Tours. Accessed August 29, 2014. http://www.dmztours.com.

Drenthen, Martin. "Ecological Restoration and Place Attachment: Emplacing Non-places?" *Environmental Values* 18 (2009): 285–312.

Dryzek, John S. *The Politics of the Earth: Environmental Discourses*, 2nd ed. Oxford: Oxford University Press, 2005.

Dudley, Marianna. *An Environmental History of the UK Defence Estate, 1945 to the Present*. London: Bloomsbury, 2012.

Dumanoski, Dianne. "Pentagon Takes First Steps toward Tackling Pollution." *Boston Globe*, September 9, 1990.

Durant, R. F. *The Greening of the U.S. Military: Environmental Policy, National Security, and Organizational Change*. Washington, DC: Georgetown University Press, 2007.

Eckert, Astrid M. "'Greetings from the Zonal Border': Tourism to the Iron Curtain in West Germany." *Zeithistorische Forschungen/Studies in Contemporary History* 8 (2010): 9–36.

Edensor, Tim. *Industrial Ruins: Space, Aesthetics, and Materiality*. Oxford: Berg, 2005.

Edgington, Ryan H. *Range Wars: The Environmental Contest for White Sands Missile Range*. Lincoln: University of Nebraska Press, 2014.

Egan, Dave. "Authentic Ecological Restoration." *Ecological Restoration* 24, no. 4 (2006): 223–24.

Eilperin, Juliet. "Obama Proposes Vast Expansion of Pacific Ocean Sanctuaries for Marine Life." *Washington Post*, June 17, 2014. http://www.washingtonpost.com/politics/obama-will-propose-vast-expansion-of-pacific-ocean-marine-sanctuary/2014/06/16/f8689972-f0c6-11e3-bf76-447a5df6411f_story.html.

———. "Trump Orders Review of National Monuments, Vows to 'End These Abuses and Return Control to the People.'" *Washington Post*, April 26, 2017. https://www.washingtonpost.com/news/energy-environment/wp/2017/04/25/zinke-to-review-large-national-monuments-created-since-1996-to-make-sure-the-people-have-a-voice/.

Elliot, Robert. "Faking Nature." *Inquiry* 25, no. 1 (1982): 81–93.

———. *Faking Nature: The Ethics of Environmental Restoration*. London: Routledge, 1997.

Engineering News Record. "Asphaltic Membrane Is Used to Leakproof a Lake." November 22, 1956, 40–41. Rocky Mountain Arsenal JARDF document no. B5600033.

European Green Belt. "Memorial Landscape." Accessed September 4, 2015. http://www.europeangreenbelt.org/iron-curtain.html.

European Green Belt Initiative. "Borders Separate, Nature Unites!" Accessed August 26, 2014. http://www.europeangreenbelt.org/initiative.html.

Farish, Matthew. *The Contours of America's Cold War*. Minneapolis: University of Minnesota Press, 2010.

Farquhar, Brodie. "Revealed—Secret Changes to Park Rules." *High Country News*, September 19, 2005.

Finley, Bruce. "Security Water Raises Concern." *Colorado Springs Gazette*, June 16, 2016, A3–A4.

Fischman, Robert L. *The National Wildlife Refuges: Coordinating a Conservation System through Law*. Washington, DC: Island Press, 2003.

Foote, Kenneth E. *Shadowed Ground: America's Landscapes of Violence and Tragedy*. Austin: University of Texas Press, 2003.

Fox, Ben. "U.S. Rattles Puerto Rico with Bomb Site Cleanup Plan." *USA Today*, October 4, 2012. https://www.usatoday.com/story/news/world/2012/10/04/puerto-rico-bomb-cleanup-plan/1614149/.

Galatowitsch, Susan M. *Ecological Restoration*. Sunderland, MA: Sinauer, 2012.

Geidezis, Liana, and Melanie Kreutz. "Green Belt Europe—Structure of the Initiative and Significance for a Pan European Ecological Network." In *Proceedings of the 1st GreenNet Conference: The Green Belt as a European Ecological Network—Strengths and Gaps*, edited by Ilke Marschall, Marion Muller, and Matthias Gather, 12–21. GreenNet.

Glass, Stephen B., Bradley M. Herrick, and Christopher J. Kucharik. "Climate Change and Ecological Restoration at the University of Wisconsin–Madison Arboretum." *Ecological Restoration* 27 (2009): 345–49.

GlobalSecurity.org. "Piñon Canyon Maneuver Site (PCMS)." Last updated May 7, 2011. http://www.globalsecurity.org/military/facility/pinon-canyon.htm.

Gobster, Paul H. "Urban Park Restoration and the 'Museumification' of Nature." *Nature and Culture* 2 (2007): 95–114.

Gonzalez, George A. "Local Growth Coalitions and Air Pollution Controls: The Ecological Modernisation of the U.S. in Historical Perspective." *Environmental Politics* 11, no. 3 (2002): 121–44.

Goren, Lilly J. *The Politics of Military Base Closings: Not in My District*. New York: Peter Lang, 2003.

Government of Western Australia, Parks and Wildland Service. "Montebello Islands Marine Park: Explosive Attraction." Accessed August 20, 2015. http://parks.dpaw.wa.gov.au/park/montebello-islands.

Grabosky, P. N. "Chapter 16: A Toxic Legacy: British Nuclear Weapons Testing in Australia." In *Wayward Governance: Illegality and Its Control in the Public Sector* by Grabosky, 235–53, Canberra: Australian Institute of Criminology, 1989. Last modified July 29, 2009. http://www.aic.gov.au/publications/previous%20series/lcj/1-20/wayward/ch16.html.

Guy, Andrew, Jr. "'Bomblet' at Arsenal Cancels All Tours." *Denver Post*, October 21, 2000, A1.

Gyeonggi G-News. "Gyeonggi Province Adds Stories to Pyeonghwa Nuri-gil." June 2, 2014. http://gnews2824.blogspot.com/2014/06/gyeonggi-province-adds-stories-to.html (blog discontinued).

Hajer, Maarten A. "Ecological Modernisation as Cultural Politics." In *Risk, Environment and Modernity: Towards a New Ecology*, edited by Scott Lash, Bronislaw Szerszynski, and Brian Wynne, 246–68. London: Sage Publications, 1996.

———. *The Politics of Environmental Discourse: Ecological Modernization and the Policy Process*. Oxford: Clarendon, 1995.

Hamblin, Jacob Darwin. *Arming Mother Nature: The Birth of Catastrophic Environmentalism*. NY: Oxford University Press, 2013.

Hammer, Joshua. "A Barrier Gone, but Not Forgotten." *New York Times*, July 26, 2009, TR1.

Hancock, Lee. "Eagles' Don Henley Works to Preserve East Texas' Caddo Lake." *Dallas Morning News*, April 25, 2011. http://www.dallasnews.com/news/state/headlines/20110425-eagles-don-henley-works-to-preserve-east-texas-caddo-lake.ece.

Hansen, Dan. "Free-Flowing Debate Control over Hanford Reach: Should It Be Feds or Counties?" *Spokesman-Review*, June 15, 1997. http://www.spokesman.com/stories/1997/jun/15/free-flowing-debate-control-over-hanford-reach/.

Harris, Peter. "Militarism in Environmental Disguise: The Greenwashing of an Overseas Military Base." *International Political Sociology* 9, no. 1 (2015): 19–36.

Harvey, David. *Justice, Nature and the Geography of Difference*. Cambridge, MA: Blackwell, 1996.

Hautzinger, Sarah, and Jean Scandlyn. *Beyond Post-Traumatic Stress: Homefront Struggles with the Wars on Terror*. Walnut Creek, CA: Left Coast Press, 2013.

Havlick, David G. "Disarming Nature: Converting Military Lands to Wildlife Refuges." *Geographical Review* 101, no. 2 (2011): 183–200.

———. "Logics of Change for Military-to-Wildlife Conversions in the United States." *GeoJournal* 69, no. 3 (2007): 151–64.

———. "Militarization, Conservation, and U.S. Base Transformations." In *Militarized Landscapes: From Gettysburg to Salisbury Plain*, edited by C. J. Pearson, T. Cole, and P. Coates, 113–33. London: Bloomsbury, 2010.

———. "Reconsidering Wilderness: Prospective Ethics for Nature, Technology, and Society." *Ethics, Place and Environment* 9, no. 1 (2006): 47–62.

———. "Restoration, History, and Values at Transitioning Military Sites in the United States." In *Restoring Layered Landscapes: History, Ecology, Culture*, edited by Marion Hourdequin and D. G. Havlick, 160–80. New York: Oxford University Press, 2016.

Havlick, David G., Marion Hourdequin, and Matthew John. "Examining Restoration Goals at a Former Military Site." *Nature and Culture* 9, no. 3 (2014): 288–315.

Havlick, David G., and Eric Perramond. "Militarized Spaces and Open Range: Piñon Canyon and (Counter)Cartographies of Rural Resistance." *Environment and Planning D* 33, no. 1 (2015): 169–84.

Higgins, Michelle. "Affordable Caribbean: Vieques." *New York Times*, October 28, 2007. http://www.nytimes.com/2007/10/28/travel/28vieques.html.

Higuchi, Hiroyoshi, Kiyoaki Ozaki, Go Fujita, Jason Minton, Mutsuyuki Ueta, Masaki Soma, and Nagahisa Mita, "Satellite Tracking of White-Naped Crane Migration and the Importance of the Korean Demilitarized Zone." *Conservation Biology* 10, no. 3 (1996): 806–12.

Hobbs, Richard J., Eric S. Higgs, and Carol Hall. *Novel Ecosystems: Intervening in the New Ecological World Order*. New York: Wiley and Sons, 2013.

Hobbs, Richard J., Eric Higgs, and James A. Harris. "Novel Ecosystems: Implications for Conservation and Restoration." *Trends in Ecology and Evolution* 24, no. 11 (2009): 599–605.

Hoffecker, John F. *Twenty-Seven Square Miles*. Commerce City, CO: US Fish and Wildlife Service, Rocky Mountain Arsenal National Wildlife Refuge, 2001.

Holland, Alan, and John O'Neill. "Yew Trees, Butterflies, Rotting Boots and Washing Lines: The Importance of Narrative." In *Moral and Political Reasoning in Environmental Practice*, edited by Andrew Light and Avner De-Shalit, 219–35. Cambridge, MA: MIT Press, 2003.

Hoosier Environmental Council. "National Wildlife Refuge Proposal for Jefferson Proving Ground in Trouble!" Action alert. Photocopy from Big Oaks National Wildlife Refuge files. Original printed from website with access date of April 28, 1999.

Hourdequin, Marion. *Environmental Ethics: From Theory to Practice*. New York: Bloomsbury, 2015.

Hourdequin, Marion, and David G. Havlick. "Ecological Restoration in Context: Ethics and the Naturalization of Former Military Lands." *Ethics, Policy, and Environment* 14 (2011): 69–89.

———. "Restoration and Authenticity Revisited." *Environmental Ethics* 35, no. 1 (2013): 79–93.

———, eds. *Restoring Layered Landscapes: History, Ecology, Culture*. New York: Oxford University Press, 2016.

Indies Trader. "Diving Crystal Clear Water." Accessed August 31, 2015. http://www.indiestrader.com/marshall-islands/activities/diving/.

Iron Curtain Trail. "The Iron Curtain Trail: Experiencing the History of Europe's Division." Accessed December 3, 2013. http://www.ironcurtaintrail.eu/en/der_iron_curtain_trail/index.htm.

Irwin, Alan, and Brian Wynne. Introduction to *Misunderstanding Science? The Public Reconstruction of Science and Technology*, edited by Irwin and Wynne, 8–9. Cambridge: Cambridge University Press, 1996.

Iverson, Kristen. *Full Body Burden: Growing Up in the Nuclear Shadow of Rocky Flats*. New York: Broadway, 2012.

Jackson, Richard, Marsh Lavenue, and Abhishek Singh. "Review of Long-Term Monitoring Plan for Rocky Mountain Arsenal." Report for the Site Specific Advisory Board of the Rocky Mountain Arsenal, Inc. April 8, 2011. http://www.rma.army.mil/files/1013/9214/9977/Part4_2010_FYRR_Vol._I.pdf.

Jacoby, Karl. *Crimes against Nature: Squatters, Poachers, Thieves and the Hidden History of American Conservation*. Berkeley: University of California Press, 2001.

Jungyun, Kwon. "Korea's DMZ, an Oasis of Wildlife." Homestay Korea, September 26, 2012. http://www.homestaykorea.com/?document_srl=58283.

Juniper, Chris. "Fort Carson's Extraordinary Sustainability Journey." September 21, 2007. Available at https://www.yumpu.com/en/document/view/4894280/sustainable-actions-usgbc-colorado-chapter.

———. "Fort Carson, the Mountain Post Sustainability and Environmental Management System (SEMS)." June 17, 2008. http://www.fedcenter.gov/_kd/Items/actions.cfm?action=Show&item_id=10217&destination=ShowItem.

Kanouse, Sarah. *Around Crab Orchard*. DVD, 69 mins. 2012.

Kelly, Cynthia C. "The Making of the Manhattan Project Park." *Federation of American Scientists* 68, no. 1 (Winter 2015). http://fas.org/pir-pubs/making-manhattan-project-park/.

Kim, Ke Chung. "Preserving Biodiversity in Korea's Demilitarized Zone." *Science* 278, no. 5336 (October 10, 1997): 242–43.

Kim, Suk-Young. *DMZ Crossing: Performing Emotional Citizenship along the Korean Border*. New York: University of Columbia Press, 2014.

Korea DMZ Peace-Life Valley. "Greetings." Accessed August 18, 2015. http://dmzecopeace.com/eng/01_intro/intro01.php.

Korea.net News. Accessed August 18, 2015. http://www.korea.net/NewsFocus/Society/view?articleId=100703 (webpage discontinued).

Korea Tourism Organization. "Demilitarized Zone Tours, from DMZ to PLZ." Visit Korea. Accessed August 18, 2015. http://english.visitkorea.or.kr/enu/SI/SI_EN_3_4_1.jsp.

Krupar, Shiloh. *Hot Spotter's Report: Military Fables of Toxic Waste*. Minneapolis: University of Minnesota Press, 2013.

Kuletz, Valerie L. *The Tainted Desert: Environmental Ruin in the American West*. New York: Routledge, 1998.

Kupková, Lucie, Ivan Bičík, and Jiří Najman. "Land Cover Changes along the Iron Curtain, 1990–2006." *Geografie* 118, no. 2 (2013): 95–115.

Kwon, Miwon. *One Place after Another: Site-Specific Art and Locational Identity*. Cambridge, MA: MIT Press, 2002.

Lamarra, Paul. "Riding the Iron Curtain." *Adventure Cyclist*, June 2012, 21–27.

Langhelle, Oluf. "Why Ecological Modernization and Sustainable Development Should Not Be Conflated." *Journal of Environmental Policy and Planning* 2 (2000): 303–22.

Langston, N. "Restoration in the American National Forests: Ecological Processes and Cultural Landscapes." In *The Conservation of Cultural Landscapes*, edited by Mauro Agnoletti, 163–73. Oxfordshire, England: CAB International, 2006.

Lapham, Lewis. "Lionhearts." *Harper's*, September 2006, 9–11.

Latour, Bruno. *We Have Never Been Modern*. Cambridge, MA: Harvard University Press, 1993.

Lee, Seung-ho. "A New Paradigm for Trust-Building on the Korean Peninsula: Turning Korea's DMZ into a UNESCO World Heritage Site." *Asia-Pacific Journal* 35 (2010): 2–10.

Leonard, Zeb. "Tampering with History: Varied Understanding of Operation Mosaic." *Journal of Australian Studies* 38, no. 2 (2014): 205–19.

Leopold, Aldo. *A Sand County Almanac and Sketches Here and There*. Oxford: Oxford University Press, (1949) 1987.

Leslie, Michele, Gary K. Meffe, and Jeffrey L. Hardesty. *Conserving Biodiversity on Military Lands: A Handbook for Natural Resource Managers*. Washington, DC: Department of Defense Biodiversity Initiative, US Department of Defense, and Nature Conservancy, 1996.

Light, Andrew. "Ecological Restoration and the Culture of Nature: A Pragmatic Perspective." In *Readings in the Philosophy of Technology*, 2nd ed., edited by David M. Kaplan, 452–67. Lanham, MD: Rowman and Littlefield, 2009.

Lillie, Thomas H., and J. Douglas Ripley. "A Strategy for Implementing Ecosystem Management in the United States Air Force." *Natural Areas Journal* 18, no. 1 (1998): 73–80.

Louden, Tamara. "Army Expansion Plans Have Ranchers on Edge." *Denver Post*, August 13 2006, 1E–2E.

Löwenheim, Oded. *The Politics of the Trail: Reflexive Mountain Biking along the Frontier of Jerusalem*. Ann Arbor: University of Michigan Press, 2014.

Lust4Rust. "Bikini Atoll." Accessed June 26, 2016. http://www.petemesley.com/lust4rust/wreck-trips/bikini-atoll/.

Lutz, Catherine, ed. *The Bases of Empire: The Global Struggle against U.S. Military Bases*. New York: New York University Press, 2008.

Machlis, Gary E., and Thor Hanson. "Warfare Ecology." *BioScience* 58 (September 2008): 729–36.

Maclellan, Nic. "Cleaning Up Johnston Atoll." Nautilus Institute. APSNet Special Reports, November 25, 2005. http://nautilus.org/apsnet/cleaning-up-johnston-atoll/.

Marshall Islands Dose Assessment & Radioecology Program. "Bikini Atoll." Last modified April 7, 2015. https://marshallislands.llnl.gov/bikini.php.

McCaffrey, Katherine. "Fish, Wildlife, and Bombs: The Struggle to Clean Up Vieques." North American Congress on Latin America. September/October 2009. https://nacla.org/article/fish-wildlife-and-bombs-struggle-clean-vieques.

———. *Military Power and Popular Protest: The U.S. Navy in Vieques, Puerto Rico*. New Brunswick, NJ: Rutgers University Press, 2002.

McClelland, James Robert. *The Report of the Royal Commission into British Nuclear Tests in Australia*. Canberra: Australian Government Publishing Service, 1985.

McKibben, Bill. *The End of Nature*. New York: Random House, 1989.

McMenemy, Jeff. "EPA Orders Air Force to Treat Contaminated Wells at Pease." July 9, 2015. http://www.seacoastonline.com/article/20150709/NEWS/150709082.

Meierotto, Lisa. "A Disciplined Space: The Co-evolution of Conservation and Militarization on the US-Mexico Border." *Anthropological Quarterly* 87, no. 3 (Summer 2014): 637–64.

"Memorandum of Agreement between US Army Test and Evaluation Command and Region 3, US Fish and Wildlife Service for Natural Resource Management of the Firing Range of the Jefferson Proving Ground." May 5, 1997. Photocopy in Big Oaks National Wildlife Refuge files.

Meyerson, Harvey. *Nature's Army: When Soldiers Fought for Yosemite*. Lawrence: University of Kansas Press, 2001.

Misrach, Richard, with Myriam Weisang Misrach. *Bravo 20: The Bombing of the American West*. Baltimore: Johns Hopkins University Press, 1990.

Mitchell, Don. *The Lie of the Land: Migrant Workers and the California Landscape*. Minneapolis: University of Minnesota Press, 1996.

Mol, Arthur P. J. "Ecological Modernisation and Institutional Reflexivity: Environmental Reform in the Late Modern Age." *Environmental Politics* 5 (1996): 302–23.

———. "The Environmental Movement in an Era of Ecological Modernisation." *Geoforum* 31 (2000): 45–56.

Mol, Arthur P. J., and D. Sonnenfeld, "Ecological Modernisation around the World: An Introduction." *Environmental Politics* 9, no. 1 (2000): 3–16.

Mol, Arthur P. J., and Gert Spaargaren, "Ecological Modernisation Theory in Debate: A Review." *Environmental Politics* 9, no. 1 (2000): 17–49.

———. "Environment, Modernity, and the Risk-Society: The Apocalyptic Horizon of Environmental Reform." *International Sociology* 8, no. 4 (1993): 431–59.

Montebello Islands Marine Park. "Explosive Attraction." Accessed August 20, 2015. http://parks.dpaw.wa.gov.au/park/montebello-islands.

Murray, Patty. "Senator Murray Applauds Hanford Reach Designation" (press release). June 9, 2000. http://www.murray.senate.gov/public/index.cfm/2000/6/senator-murray-applauds-hanford-reach-designation.

Mythen, Gabe. *Ulrich Beck: A Critical Introduction to the Risk Society*. London: Pluto Press, 2004.

National Park Service. "Draft Management Policies to Guide the National Park Service." Department of the Interior, June 16, 2006.

National Park Service, Gettysburg National Military Park. "Management." Accessed January 7, 2016. http://www.nps.gov/gett/learn/management/index.htm.

National Park Service, Little Bighorn Battlefield. "Little Bighorn Battlefield National Monument Resources Management Plan." July 9, 2007. http://www.nps.gov/libi/learn/management/upload/ResourceManagementPlan.pdf.

National Park Service, Manhattan Project National Historical Park. "Learn about the Park." Accessed June 21, 2016. https://www.nps.gov/mapr/learn/index.htm.

National Park Service, Trail of Tears National Historic Trail. "A Journey of Injustice." Last updated December 21, 2016. http://www.nps.gov/trte/index.htm.

National Renewable Energy Laboratory. "Helping Fort Carson Meet DOD Energy Goals." *Continuum* 7 (Fall 2014). http://www.nrel.gov/continuum/partnering/defense.html.

National TLC Service. September 2011. http://www.readysubjects.org/nationaltlc/wp-content/uploads/2011/11/brochure-3panel_final.pdf. Brochure.

National Toxic Land/Labor Conservation Service. Accessed June 13, 2016. http://www.nationaltlcservice.us/.

National Trust. "Orford Ness National Nature Reserve." Accessed June 9, 2016. https://www.nationaltrust.org.uk/orford-ness-national-nature-reserve.

———. "Orford Ness, 2009." http://www.tourismleafletsonline.com/pdfs/NT_%20Orford_Ness_Leaflet.pdf. Brochure.

Nature Conservancy. "Our World: 2015 Annual Report." https://www.nature.org/media/annualreport/2015-annual-report.pdf.

Navarro, Mireya. "Navy's Vieques Training May Be Tied to Health Risks." *New York Times*. November 14, 2009, A14.

———. "New Battle on Vieques, over Navy's Cleanup of Munitions." *New York Times*, August 6, 2009, A10.

Nazaryan, Alexander. "Tiptoeing through America's Sarin Stash." *Newsweek*, October 11, 2013. http://www.newsweek.com/2013/10/11/tiptoeing-through-americas-sarin-stash-238108.html.

Nguyen, Viet Thanh. *Nothing Ever Dies: Vietnam and the Memory of War*. Cambridge, MA: Harvard University Press, 2016.

Niedenthal, Jack. "A History of the People of Bikini Following Nuclear Weapons Testing in the Marshall Islands: With Recollections and Views of Elders of Bikini Atoll." *Health Physics* 73, no. 1 (1997): 28–36.

Office of Budget and Management Department of Defense budget, FY 2007. Accessed August 31, 2006. https://www.whitehouse.gov/omb/budget/fy2007/pdf/budget/defense.pdf (webpage discontinued).

Office of the Undersecretary of Defense. "Operation and Maintenance Overview, Fiscal Year 2016 Budget Estimates." February 2015. http://comptroller.defense.gov/Portals/45/Documents/defbudget/fy2016/fy2016_OM_Overview.pdf.

Olsen, Ken. "At Hanford, the Real Estate Is Hot." *High Country News*, January 22, 1996, 1, 8–9.

Omaha (Nebraska) *World-Herald*. "Leakproof Bottom Underlies 100-Acre Lake Near Denver." November 11, 1956. Rocky Mountain Arsenal JARDF document no. B5600032.

Palka, Eugene J., and Francis A. Galgano. *The Scope of Military Geography: Across the Spectrum from Peacetime to War*. New York: McGraw-Hill, 2000.

Pearson, Chris, Peter Coates, and Tim Cole, eds. *Militarized Landscapes: From Gettysburg to Salisbury Plain*. London: Bloomsbury, 2010.

Pearson, James, and Paul Ingrassia. "Tank Traps Meet Tourist Traps at Korea's Demilitarized Zone." Reuters, October 25, 2013. http://www.reuters.com/article/us-korea-dmz-tourism-idUSBRE99O0SP20131025.

Pew Project on National Security, Energy, and Climate. "Reenergizing America's Defense." Washington, DC: Pew Charitable Trusts, 2010.

Philipps, David. *Lethal Warriors: When the Band of Brothers Came Home*. New York: St. Martin's, 2010.

———. "Veterans Feel Cost of U.S. Nuclear Tests." *New York Times*, January 29, 2017, A1.

Piñon Canyon Expansion Opposition Coalition. "Our Heritage, Our History." Kim, Colorado. N.d. Brochure.

Platt, John R. "South Korea Seeks to Protect Endangered Species in Demilitarized Zone." *Scientific American*, September 27, 2011. http://blogs.scientificamerican.com/extinction-countdown/south-korea-seeks-rotect-endangered-species-demilitarized-zone/.

———. "To the Bat Bunker!" *Conservation Magazine*, September 9, 2013. http://conservation magazine.org/2013/09/bat-bunker/.

Preparatory Commission for the Comprehensive Nuclear-Test-Ban Treaty Organization. "The United Kingdom's Nuclear Testing Programme." Accessed August 21, 2015. https://www.ctbto.org/nuclear-testing/the-effects-of-nuclear-testing/the-united-kingdoms nuclear-testing-programme/.

Proclamation 8803—Establishment of the Fort Ord National Monument by the President of the United States of America. April 20, 2012. https://www.blm.gov/nlcs_web/sites/style/medialib/blm/ca/nlcs/Fort_Ord_NM/docs.Par.51266.File.dat/Fort%20Ord%20 Proclamation.pdf.

Reuben, Suzanne H. "Reducing Environmental Cancer Risk, 2008–2009 Annual Report, President's Cancer Panel." US Department of Health and Human Service, National Institutes of Health, National Cancer Institute, 2010.

Rich, Motoko. "What about Us, Nagasaki Asks, as Obama's Hiroshima Trip Nears." *New York Times*, May 24, 2016, A6.

Ripley, J. D., and M. Leslie. "Conserving Biodiversity on Military Lands." *Federal Facilities Environmental Journal*, Summer 1997, 93–105.

Rocky Mountain Arsenal National Wildlife Refuge Act of 1991. Joint Hearing before the Fisheries and Wildlife Conservation and the Environment Subcommittees of the Committee on Merchant Marine and Fisheries [serial no. 102-61] and the Military Installations and Facilities Subcommittee of the Committee on Armed Services, House of Representatives, 102nd Congress, 1st Session; PL 102-402. September 9, 1991.

Rocky Mountain Arsenal Remediation Venture Office. RVO fact sheet. Commerce City, Colorado. Accessed June 22, 2016. http://www.rma.army.mil/files/5714/1037/9699/RVOFACTSHEETRevise.pdf.

Roper, P. "Fort Carson Lands Copter Brigade." *Pueblo Chieftain*, March 29, 2011. http://www.chieftain.com/news/local/fort-carson-lands-copter-brigade/article_a0c365ba-59bd-11e0-be49-001cc4c002e0.html.

Rothbaum, Rebecca. "Wildlife Lives on Shawangunk Grasslands." *Poughkeepsie Journal*, June 20, 2002. http://cityguide.pojonews.com/fe/DayTrips/stories/dt_shawangunk_grasslands.asp (webpage discontinued).

Sanders, Barry. *Green Zone: The Environmental Costs of Militarism*. Oakland, CA: AK Press, 2009.

Schmidt, William E. "Nature Sows Life Where Man Brewed Death." *New York Times*, March 12, 1989, sec. 1, p. 1.

Schwägerl, Christian. "Along Scar from Iron Curtain, a Green Belt Rises in Germany." *Yale Environment 360*. April 4, 2011. http://e360.yale.edu/feature/along_scar_from_iron_curtain_a_green_belt_rises_in_germany/2390/.

Sellars, Richard West. *Preserving Nature in the National Parks*. New Haven, CT: Yale University Press, 1997.

Shields, Deirdre. "Defence of Nature's Realm: Soldiers Are Not the Only Creatures Who Enjoy the Army Training Grounds." *Field* (1996): 92–95.

Shin, Youngsun. "A Critical Review of Peace Park Tourism Planning and Development in the Border Region." *Tourism and Hospitality Planning and Development* 4, no. 2 (2007): 111–20.

Shulman, Seth. *The Threat at Home: Confronting the Toxic Legacy of the U.S. Military*. Boston, MA: Beacon Press, 1992.

Sierra Club Foundation. Annual report, 2014. https://www.sierraclubfoundation.org/sites/sierraclubfoundation.org/files/uploads/TSCF%20Annual%20Report%202014.pdf.

Smale, Alison. "Migrants Race North as Hungary Builds a Fence." *New York Times*, August 25, 2015, A1.

Smith, Allan E., and William E. Moore. "Report of the Radiological Clean-Up of Bikini Atoll." Washington, DC: US Atomic Energy Commission, 1972.

Society for Ecological Restoration, International Science & Policy Working Group. "Mission and Vision." Accessed January 23, 2017. http://www.ser.org/page/MissionandVision.

———. "The SER International Primer on Ecological Restoration," version 2. Tucson, AZ: Society for Ecological Restoration International, 2004. http://c.ymcdn.com/sites/www.ser.org/resource/resmgr/custompages/publications/SER_Primer/ser_primer.pdf.

Sola, Katie. "Germany to Turn 62 Military Bases into Nature Sanctuaries for Birds, Beetles, and Bats." *Huffington Post*, June 19, 2015. http://www.huffingtonpost.com/2015/06/19/german-military-bases-nature-reserves_n_7623882.html.

Sorenson, David S. *Shutting Down the Cold War: The Politics of Military Base Closure*. New York: St. Martin's, 1998.

Steffan, Will, Paul J. Crutzen, and John R. McNeill. "The Anthropocene: Are Humans Now Overwhelming the Great Forces of Nature?" *AMBIO* 36, no. 8 (2007): 614–21.

Sullivan, Kevin. "Borderline Absurdity: A Fun-Filled Tour of the Korean DMZ." *Washington Post*, January 11, 1998, E1.

Šumava National Park. "Nature: Basic Information." Accessed June 22, 2016. http://www.npsumava.cz/en/3261/sekce/basic-information/.

Tangley, Laura. "Bases Loaded." *National Wildlife* 43 (October/November 2005): 38–45.

Tetra Tech EC, Inc., for Rocky Mountain Arsenal Remediation Venture Office. "Rocky Mountain Arsenal: Final 2010 Five-Year Review Report for Rocky Mountain Arsenal, Commerce City, Adams County, Colorado, Review Period: April 1, 2005–March 31, 2010." September 2011. Available at https://www.epa.gov/sites/production/files/documents/RMA_2010FYRVol-I_9-7-2011.pdf.

Thomas, Julia Adeney. "The Exquisite Corpses of Nature and History: The Case of the Korean DMZ." *Asia-Pacific Journal* 7, no. 43 (2009): 1–16.

Tierney, John R. "Case Study of the Establishment of Great Bay National Wildlife Refuge at the Former Pease Air Force Base, New Hampshire." Master's thesis, Natural Resources and Environmental Conservation, University of New Hampshire, Durham, 2001.

2005 Defense Base Closure and Realignment Commission. Final report. Arlington, VA, September 8, 2005. http://www.brac.gov/docs/final/Volume1BRACReport.pdf.

UNESCO (United Nations Educational, Scientific, and Cultural Organization). "Bikini Atoll Nuclear Test Site." Accessed July 3, 2017. http://whc.unesco.org/en/list/1339.

US Air Force. "Energy Strategic Plan to a Net Zero Installation." Prepared by 10th Air Base Wing and USAFA Dean of Faculty, US Air Force Academy Headquarters, Colorado Springs, Colorado, March 2009.

———. "U.S. Air Force Energy, Environment, Safety and Occupational Health: Managing for Operational Sustainability, 2007 Inaugural Report." Washington, DC, 2007.

US Department of the Army. "Army Earth Day Message 2010." April 8, 2010. https://www.army.mil/e2/-images/2010/04/08/69329/index.html.

———. "Army Strategy for the Environment: Sustain the Mission, Secure the Future." Washington, DC, 2004.

———. "Earth Day 2016: Army Message: Acknowledge the Past, Engage the Present, and Chart the Future." Accessed July 8, 2017. https://www.army.mil/e2/c/downloads/430546.pdf.

———. "Environmental Protection and Enhancement." Pamphlet 200-1. January 17, 2002.

———. "Fort Carson 2014 Sustainability Report." Fort Carson, Colorado, 2015.

———. "Jefferson Proving Ground Final Environmental Impact Statement: Disposal and Reuse of the Jefferson Proving Ground." Madison, Indiana, 1995.

———. "Sustainability Report 2014." Office of the Assistant Secretary of the Army for Installations, Energy, and Environment. September 29, 2014. http://www.asaie.army.mil/Public/ES/doc/Army%20Sustainability%20Report%202014.pdf.

US Department of the Army, Fort Carson Public Affairs Office. "Fort Carson Statistics." 2006. http://www.carson.army.mil/pao/media_relations/data_card.htm (webpage discontinued).

US Department of Defense. "Base Realignment and Closures: Report of the Defense Secretary's Commission." December 1988. http://www.acq.osd.mil/brac/Downloads/Prior%20BRAC%20Rounds/1988.pdf.

———. "Base Structure Report: Fiscal Year 2003 Baseline." June 2003. http://archive.defense.gov/news/Jun2003/basestructure2003.pdf.

———. "Defense Environmental Programs: Annual Report to Congress for FY 2013." June 2014. http://www.denix.osd.mil/arc/arcfy2013/report/fiscal-year-2013-defense-environmental-programs-annual-report-to-congress/.

———. "Defense Environmental Programs Annual Report to Congress for FY 2014." September 2015. http://www.denix.osd.mil/arc/arcfy2014/report/fiscal-year-2014-defense-environmental-programs-annual-report-to-congress/.

US Department of Energy. "Stewards of National Resources." DOE/FM-0002. 1994.

US Department of Energy, Office of Civilian Radioactive Waste Management. "Final Environmental Impact Statement for a Geologic Repository for the Disposal of Spent Nuclear Fuel and High-Level Radioactive Waste at Yucca Mountain, Nye County, Nevada. Readers Guide and Summary." DOE/EIS-0250. February 2002. https://energy.gov/sites/prod/files/EIS-0250-FEIS_Summary-2002.pdf.

US Department of Health and Human Services. "An Evaluation of Environmental, Biological, and Health Data from the Island of Vieques, Puerto Rico." Agency for Toxic Substances and Disease Registry, Division of Community Health Investigations, Atlanta, Georgia, March 19, 2013.

US Department of the Interior. "Budget Justifications and Performance Indication, Fiscal Year 2016: US Fish and Wildlife Service." http://www.fws.gov/budget/2015/FY2016_FWS_Greenbook.pdf.

US Environmental Protection Agency. "The Environmental Challenge of Military Munitions and Federal Facilities." Last updated, June 24, 2015. https://www.epa.gov/enforcement/environmental-challenge-military-munitions-and-federal-facilities.

———. "Superfund: Site Information for Atlantic Fleet Weapons Training Area." December 14, 2015. http://cumulis.epa.gov/supercpad/cursites/dsp_ssppSiteData1.cfm?id=0204694#Status.

US Fish and Wildlife Service. "Assabet River National Wildlife Refuge." Sudbury, MA, May 2010. Brochure.

———. "Budget Justifications and Performance Information, Fiscal Year 2013." US Department of the Interior, 2012.

———. "Budget Justifications and Performance Information, Fiscal Year 2015." US Department of the Interior, 2014. http://www.fws.gov/budget/2014/FY2015_FWS_Greenbook-DOI31014.pdf.

———. "Eastern Massachusetts National Wildlife Refuge Complex." Sudbury, MA, June 2001. Brochure.

———. "Former Bombing Range Becomes National Wildlife Refuge." *Fish and Wildlife News*, July/August 2000. http://news.fws.gov/articles/FormerBombing.html.

———. "President Obama Protects Vital Marine Habitat, Expands Pacific Remote Islands Marine National Monument." September 25, 2014. https://www.fws.gov/pacific/news/news.cfm?id=2144375346.

———. "Retired Military Bunkers Used as Artificial Bat Hibernacula at Aroostook National Wildlife Refuge in Maine." News release, April 9, 2013. Last updated May 10, 2016. https://www.fws.gov/news/ShowNews.cfm?ID=0DA476C9-AE33-F4D0-2BE4EF9E2ECBF61B.

———. "Rocky Mountain Arsenal National Wildlife Refuge: A Place like No Other." Department of the Interior, 1999, Commerce City, Colorado. Brochure.

———. "Vieques National Wildlife Refuge: Comprehensive Conservation Plan and Environmental Impact Statement." USDI-FWS Southeast Region, August 2007. Available at https://ecos.fws.gov/ServCat/DownloadFile/18724?Reference=1473.

US Fish and Wildlife Service, Aroostook National Wildlife Refuge. "Auto Tour Route." Last updated December 13, 2013. https://www.fws.gov/refuge/aroostook/visit/autotour.html.

———. "Plan Your Visit." Last updated September 20, 2016. https://www.fws.gov/refuge/Aroostook/visit/plan_your_visit.html.

US Fish and Wildlife Service, Big Oaks National Wildlife Refuge. "About the Refuge." Last updated May 28, 2016. https://www.fws.gov/refuge/Big_Oaks/about.html.

———. "Fire Management Plan, Environmental Assessment: Appendix K." Madison, IN, March 2001.

US Fish and Wildlife Service, Caddo Lake National Wildlife Refuge. "History of the Longhorn Army Ammunition Plant." Last updated August 17, 2012. https://www.fws.gov/refuge/Caddo_Lake/about/amm_plant.html.

———. "Paddlefish Restoration and Recovery." Last updated May 21, 2014. https://www.fws.gov/refuge/caddo_lake/wildlife/paddlefish.html.

———. "Plan Your Visit." Last updated August 14, 2012. https://www.fws.gov/refuge/Caddo_Lake/visit/plan_your_visit.html.

———. "Wildlife & Habitat." Last updated January 3, 2013. https://www.fws.gov/refuge/Caddo_Lake/wildlife_and_habitat/index.html.

US Fish and Wildlife Service, Crab Orchard National Wildlife Refuge. "Comprehensive Conservation Plan." Fort Snelling, Minnesota, November 2006.

———. "Draft Environmental Impact Statement/Comprehensive Conservation Plan." Fort Snelling, Minnesota, September 2005.

US Fish and Wildlife Service, Fisherman Island National Wildlife Refuge. "Final Comprehensive Conservation Plan." Last updated April 1, 2013. http://www.fws.gov/refuge/Fisherman_Island/what_we_do/finalccp.html.

———. "Refuge Tour." Last updated July 11, 2014. http://www.fws.gov/refuge/Fisherman_Island/events/refugetour.html.

US Fish and Wildlife Service, Fort Niobrara National Wildlife Refuge. "About the Refuge." Last updated February 21, 2014. https://www.fws.gov/refuge/Fort_Niobrara/about.html.

US Fish and Wildlife Service, Great Bay National Wildlife Refuge. "About the Refuge." Last updated November 14, 2014. http://www.fws.gov/refuge/Great_Bay/about.html.

US Fish and Wildlife Service, Hanford Reach National Wildlife Refuge. "Habitats." May 3, 2013. https://www.fws.gov/refuge/Hanford_Reach/Wildlife_Habitat/Habitat.html.

US Fish and Wildlife Service, National Wildlife Refuge System. "About: Mission." Last updated October 15, 2015. http://www.fws.gov/refuges/about/mission.html.

———. *Fulfilling the Promise: The National Wildlife Refuge System*. US Department of the Interior, March 1999.

———. "National Wildlife Refuge System Improvement Act of 1997: Public Law 105-57." Last updated August 19, 2009. http://www.fws.gov/refuges/policiesandbudget/hr1420_index.html.

US Fish and Wildlife Service, Nomans Land Island National Wildlife Refuge. "About the Refuge." Last updated October 19, 2011. https://www.fws.gov/refuge/nomans_land_island/.

US Fish and Wildlife Service, Rocky Mountain Arsenal National Wildlife Refuge. "About the Refuge." Last updated November 5, 2015. http://www.fws.gov/refuge/Rocky_Mountain_Arsenal/about.html.

———. "Resource Management." Last updated March 7, 2013. http://www.fws.gov/refuge/Rocky_Mountain_Arsenal/what_we_do/resource_management.html.

US Fish and Wildlife Service, Shawangunk Grasslands National Wildlife Refuge. "History." Last updated May 27, 2015. http://www.fws.gov/northeast/shawangunk/history.htm.

———. "Visitor Opportunities." Last updated January 26, 2012. http://www.fws.gov/northeast/shawangunk/visitor%20opportunities.htm.

US Government Accountability Office. "Defense Infrastructure: DOD Efforts Regarding Net Zero Goals." GAO-16-153R. January 12, 2016. http://www.gao.gov/assets/680/674599.pdf.

US Government Accounting Office. "Military Base Closures: Overview of Economic Recovery, Property Transfer, and Environmental Cleanup." GAO-01-1054T. Released August 28, 2001.

———. "National Wildlife Refuges: Continuing Problems with Incompatible Uses Call for Bold Action." GAO/RCED-89-196. 1989.

US House of Representatives, Committee on Natural Resources. "Buying More Land When We Can't Maintain What We Already Own: The National Wildlife Refuge System's Operations and Maintenance Backlog Story!" Oversight hearing before the Subcommittee on Fisheries, Wildlife, Oceans and Insular Affairs, 112th Congress, 1st Session, May 26, 2011. Serial no. 112-35. https://www.gpo.gov/fdsys/pkg/CHRG-112hhrg66651/pdf/CHRG-112hhrg66651.pdf.

US Navy, Naval Facilities Engineering Command. "Former Atlantic Fleet Weapons Training Area—Vieques." Last updated September 7, 2015. http://www.navfac.navy.mil/products_and_services/ev/products_and_services/env_restoration/installation_map/navfac_atlantic/vieques.html.

US Nuclear Regulatory Commission. "Jefferson Proving Ground." Last reviewed/updated October 20, 2016. https://www.nrc.gov/info-finder/decommissioning/complex/jefferson-proving-ground-facility.html.

US White House. "Establishment of the Hanford Reach National Monument, A Proclamation." June 9, 2000. https://clintonwhitehouse4.archives.gov/CEQ/hanford_reach_proclamation.html.

———. "Fact Sheet: President Obama to Designate Largest Marine Monument in the World Off-Limits to Development." Office of the Press Secretary, September 24, 2014. https://www.whitehouse.gov/the-press-office/2014/09/24/fact-sheet-president-obama-designate-largest-marine-monument-world-limit.

Verheyen, Dirk. *United City, Divided Memories? Cold War Legacies in Contemporary Berlin*. Lanham, MD: Lexington Books, 2008.

Vine, David. 2015. *Base Nation: How U.S. Military Bases Abroad Harm America and the World*. New York: Henry Holt.

Vine, David, and Miriam Pemberton. "Marine Protection as Empire Expansion." *Foreign Policy in Focus*, May 6, 2009. http://fpif.org/marine_protection_as_empire_expansion/.

Wall, Tim. "Korean DMZ Teems with Wildlife." D News, February 17, 2012. http://news.discovery.com/earth/korean-dmz-teems-with-wildlife-120217.htm.

Ward, Carlton, Jr. "Bombing Range Is National Example for Wildlife Conservation." *National Geographic*, April 1, 2015. http://voices.nationalgeographic.com/2015/04/01/bombing-range-is-national-example-for-wildlife-conservation/.

Wargo, John. *Green Intelligence: Creating Environments that Protect Human Health*. New Haven, CT: Yale University Press, 2009.

Weeks, Jennifer. "From Bombs to Birds." *Defenders Magazine*, Winter 2009, 20–23.

Weisman, Alan. *The World without Us*. New York: St. Martin's, 2007.

Westing, Arthur H. "The Korean Demilitarized Zone (DMZ) as a Bridge between the Two Koreas." Accessed January 27, 2017. http://culturaldiplomacy.org/academy/content/pdf/participant-papers/2010www/The_Korean_Demilitarized_Zone_(DMZ)_as_a_bridge_between_the_two_Koreas.pdf.

———. "Toward Environmental Sustainability and Reduced Tensions on the Korean Peninsula," *Environment* 52, no. 1 (2010): 20–23.

Whatmore, Sarah. 2002. *Hybrid Geographies: Natures Cultures Spaces*. London: Sage.

White Sands Missile Range. "Trinity Site Open House." Last updated October 7, 2015. http://www.wsmr.army.mil/PAO/Trinity/Pages/Home.aspx.

Woodward, Rachel. "Khaki Conservation: An Examination of Military Environmentalist Discourses in the British Army." *Journal of Rural Studies* 17, no. 2 (2001): 201–17.

———. *Military Geographies*. Malden, MA: Blackwell, 2004.

Woodworth, Paddy. *Our Once and Future Planet: Restoring the World in the Climate Change Century*. Chicago: University of Chicago Press, 2013.

Yoneyama, Lisa. *Hiroshima Traces: Time, Space, and the Dialectics of Memory*. Berkeley: University of California Press, 1999.

———. "Memories in Ruins: Hiroshima's Nuclear Annihilation and Beyond." In *Cities into Battlefields: Metropolitan Scenarios, Experiences, and Commemorations of Total War*, edited by Stefan Goebel and Derek Keene, 185–202. Burlington, VT: Ashgate, 2011.

Zalasiewicz, Jan, Mark Williams, Alan Haywood, and Michael Ellis. "The Anthropocene: A New Epoch of Geological Time?" *Philosophical Transactions of the Royal Society A* 369 (2011): 835–41.

INDEX